BASIC STRUCTURAL
ENGINEERING PRINCIPLES

for

Construction Managers, Architects

Quantity Surveyors and Building Surveyors

First Published 2011
This Revised Edition 2017 (Changes to R.C. Design)
Available From www.lulu.com
Email: Kevan.Heathcote@gmail.com
Copyright Kevan Heathcote 2011
19 Mathews Street
Davidson
NSW
Australia
2085

ISBN 978-1-4709-7444-2

PREFACE

Construction Managers, Building Surveyors, Architects and Quantity Surveyors are not Engineers. They should however have some idea of the magnitude of loads imposed on structures, how the structures carry those loads down to the soil through the footings and how to read engineering drawings if they are to be effective members of project management teams.

In residential work non-engineering construction professionals may be required at times to use simplified structural standards such as the Australian Timber Framing Code or the Australian Residential Slabs and Footings Code. Knowledge of the structural principles behind these codes and their limitations will assist in the proper use of such codes. An understanding of structural systems can also help in the preliminary estimation of costs for a project and enable non-engineers to participate more fully in the design process.

This book seeks to introduce non-engineering construction professionals to the principles of structural design from the initial calculation of loads, to the calculation of the internal actions in members resulting from such loads and finally to a comparison between those internal actions and the member capacities. The design process will be illustrated with reference particularly to timber design but the design of reinforced concrete, prestressed concrete, steel, brick and glass is also presented in a simplified approach. The member capacity calculations given in this book are not always exact calculations according to the relevant codes of practice. This has been done where approximate methods devised by the author give reasonable results and explain the basic principles. They are not to be used for final design purposes when all calculations should be carried out by a registered Structural Engineer.

INDEX

1.   DEAD AND LIVE LOADS ON STRUCTURES

1.1   Introduction

The primary function of any structure is to transfer the loads imposed on it to the ground. Such loads can be the structures own gravitational load (dead load) or superimposed load due to people and machinery (live loads). Chapter 3 will deal with other superimposed loads such as wind, soil, earthquakes etc.
In order to analyse the effect of dead and live loads on a structure we need to be able to calculate their magnitude. However before we do this we need to know a bit about force, pressure and stress.

1.2   Force

Forces are the primary consideration in structural safety. Forces that are applied to a structure may cause failure and therefore the first consideration in any structural analysis is to determine the forces acting on a structure. Forces have a magnitude and a direction and are generally expressed in units of kiloNewtons (kN), kilo meaning 1000 and Newtons (N) being the primary unit of force. A typical reference force is the gravitational force (1 kN) exerted by a "front row forward" having a mass of around 100 kg.

Where a force is concentrated it is indicated by an arrow with the magnitude, eg for a beam with a load of 1 kN we would represent it as follows

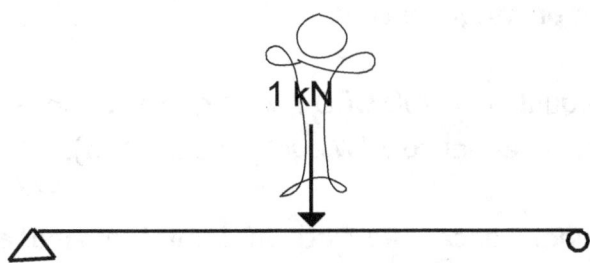

Where a force is distributed along a beam it's magnitude is indicated in kN/m as shown below. Note carefully that "w" represents force/metre and "W" represents total force.

Total Force =W= 1 kN×4 = 4kN

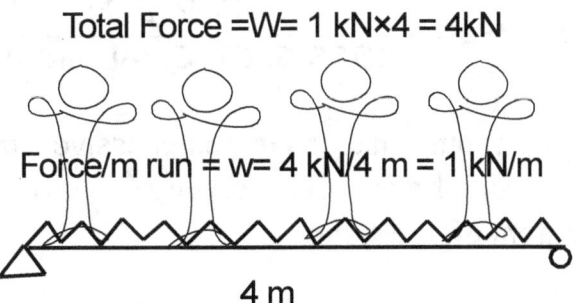

Force/m run = w= 4 kN/4 m = 1 kN/m

4 m

When a force is distributed over a surface area (eg wind force) the intensity of force is referred to as "pressure", with units of kN/m^2.

1.3 Pressure

Pressure is the term used for to express the intensity of a force . Its units are kPa (=kN/m²). It is the force per square metre of surface area and usually refers to forces that are applied <u>on</u> a body (eg wind pressure) or the intensity of forces that the body applies to another surface (eg foundation pressures). By definition

$$\text{PRESSURE (kPa)} = \frac{\text{FORCE (kN)}}{\text{AREA (m}^2)}$$

OR

$$\text{FORCE (kN)} = \text{PRESSURE (kPa)} \times \text{AREA (m}^2)$$

As a reference point a good strong wind may exert a pressure of around 1 kPa on a building. Note that 1 kPa = 1 kN/m²

1.4 Stresses

Stresses are the internal reaction of a body to an applied force or pressure (If your lecturer applies pressure for you to perform (external) your body will undergo stress (internal)). Stress is also measure of the intensity of force, but this time inside the body. Stresses can be due to compressive forces (compressive stresses), tension forces (tensile stress), bending (bending stress), shear (shear stress), torsion etc. The most common stresses we deal with are tensile stress, compressive stress, bending stress and shear stress. Stresses are measured in Newtons per square mm or Megapascals.

Note that 1 N/mm² is equal to 1,000,000 N/m² and that in Australia we call 1 N/m² a "Pascal" hence 1N/mm² is equal to 1 "Mega"pascal (MPa).

Stresses are the internal force exerted on 1 mm² of material. As a reference point the stress in a toothpick with an apple resting on it would be around 1 Mpa. Typical concrete compressive failure stresses are around 30 MPa.
For compressive and tensile forces stress is calculated as follows

$$\text{STRESS (MPa)} = \frac{\text{FORCE (N)}}{\text{CROSS SECTIONAL AREA (mm}^2)}$$

Bending stresses can vary from a maximum compressive stress at the top of the beam, to zero at the centroidal axes (centre for symmetric beams) to a maximum tensile stress at the bottom of the beam.

Shear stresses also vary down the depth of the beam but are typically maximum in the centre.

1.5   Dead Loads (DL)

Dead loads are forces ("push or a pull") due to the self weight of a structure. In some cases where the dead load is not part of the load bearing structure eg toppings on slabs or permanent partitions, the dead load is referred to as a "Superimposed dead load". Dead loads can also be defined as those loads having a coefficient of variation (Standard Deviation/Mean) less than 10%.

Of all the loads imposed on a structure dead loads are the most important because they are always there. Loads such as wind loads may or may not be acting on the structure (depending on whether the wind is blowing) but dead loads never go away. They are therefore the main reason for long-term sagging of beams

The force exerted on a structure due to dead loads is caused by the gravitational attraction of the mass of the structure towards the centre of the earth. This force can be expressed as follows –

$$\text{FORCE (N)} = \frac{G \times M \times m}{R^2}$$

where
G = Universal gravitational constant $= 6.674 \times 10^{-11}\,\text{N-m}^2/\text{kg}^2$
M = Mass of the earth (kg)
m = Mass of the structure (kg)
R = Distance from structure to centre of earth (m)

R obviously varies slightly but if we take an average value and substitute values for G and M in the above equation we can simplify it down to -

$$\text{FORCE (N)} \approx 10 \times m$$

It is obvious from this equation that force is not the same as mass (which is measured in kilograms), since it is equal to the mass times 10. If the mass is measured in kilograms then the unit of force determined by the above equation is the Newton eg. if a body has a mass of 1 kilogram we say it exerts a gravitational force of 10 Newtons. Therefore to change mass in kilograms to force in Newtons simply multiply by 10 and call it Newtons. In most cases the Newton is too small a unit of force for practical purposes. It is approximately the force exerted by an apple. For construction work we generally use kiloNewtons as our unit of force (1kN = 1000 N) . This is approximately the gravitational force exerted by a 100kg front row forward.
Expressing Force in kiloNewtons the equation becomes - Force (kN) = Mass (kg) × 10 / 1000 or -

$$FORCE\ (kN) = MASS\ (kg)\ /\ 100\ or$$

$$FORCE\ (kN) = Mass\ (tonnes) \times 10$$

You can think of the mass of a structure as the number of atoms within it. Different materials have different numbers of atoms per cubic metre, this property being referred to as the *DENSITY* of the material. In practice Mass is measured in kilograms (kg) and Density is measured in kilograms per cubic metre (kg/m^3) and the equation linking the two is -

$$Mass\ (kg) = Volume\ (m^3) \times Density\ (kg/m^3)$$

If we now substitute this equation into our equation for Force we get -

$$Force\ (Newtons) = Volume(m^3) \times Density(kg/m^3) \times 10$$

Now we can define a property of materials called *UNIT WEIGHT*, which is equal to the Density of the material times 10 divided by 1000. For example the Unit Weight of water is 1000 × 10/1000 = 10. The units of Unit Weight are *KN/m^3*. Our equation for Force then becomes -

$$FORCE\ (kN) = VOLUME\ (m^3) \times UNIT\ WEIGHT(kN/m^3)...(2)$$

The Unit Weights of building materials are given in Australian Standard AS/NZS 1170.1. Values for some common building materials from that standard are given in Table 1.1–

Table 1.1   Material Unit Weights

MATERIAL	UNIT WEIGHT (kN/m^3)
Loose Sand	18
Hardwood	11 (8.5 Seasoned)
Softwood	7.8 (5.5 Seasoned)
Glass	25.5
Uncompressed F/C Sheet	14.2
Compressed F/C Sheet	17.2
Steel	76.9
Clay Bricks	19
Calcil Bricks	18
Unfilled 200 block walls	11
Concrete	24
Reinforced Concrete	25 (Approx)
Water	10

It is easier in the case of thin materials to express the force equation in terms of the area of the material instead of the volume eg a concrete slab or roof sheeting. The equation then becomes -

$$FORCE\ (kN) = AREA\ (m^2) \times FORCE/m^2\ (kN/m^2)...(3)$$

Australian Standard AS/NZS 1170.1 "Structural design actions - Permanent, imposed and other actions" lists the Force/Square metre for common building materials eg 10 mm plasterboard has a force per square metre of 0.08 kPa - to get the dead load of this material in kN simply multiply its area by 0.08. Some of the more common materials suited to this equation are given in Table 1.2-

Table 1.2   Unit Weights Expressed as Force/m^2

MATERIAL	FORCE/m^2 (kPa)
10mm Plasterboard	0.08
Terracotta Tiles	0.57
Concrete Tiles	0.53
Terracotta or Concrete Roofs including roofing members and ceiling	0.90
10mm Glass	0.255
4.5mm F/C Sheet(Uncompressed)	0.064
0.8mm Corrugated Iron	0.10
13mm Clay Tiling	0.43
20mm H/wood flooring(seasoned)	0.17
100 mm thick Reinf. Conc. Slab	2.5

Table 1.2 can also be used for materials with different thicknesses to those given in the table. For example to find the dead load in kPa of a 180 mm thick reinforced concrete slab multiply 2.5 by 180/100 (=4.5 kPa)

It is sometimes convenient in building work to calculate dead loads of areas comprising elements at regular spacings (eg floor joists or rafters) on the basis of a force/m^2 rather than as a sum of the dead loads of the individual components. In this case simply take the dead load of a one metre length of the element and divide that dead load by the spacing of the member. For example if 200 × 50 hardwood joists are spaced at 450 mm centres then -

Force of 1m length of joist   = Volume × Unit Weight
          = (0.2×0.05×1)×11
          = 0.11 kN

Force / m^2 = 0.11 / 0.45 = 0.245 say 0.25 kPa

Example: A floor consists of 150 by 50 unseasoned hardwood joists at 450 mm centres with a 10 mm plasterboard ceiling under and 25 mm seasoned hardwood flooring boards. To calculate the dead load of this floor in kPa we proceed as follows

Dead load of flooring = 25/20 × 0.17   = 0.22 kPa

Dead load of joists = (1×0.15×0.05×11)/0.45   =0.19 kPa

Dead load of ceiling   =0.08 kPa

               Total Dead Load   =0.49 kPa

1.6   Live Loads (LL)

Live Loads are the loads imposed on a structure due to people, vehicles, movable equipment etc which by their nature may at some time be removed from the structure. The important difference between dead and live loads (other than the fact that live loads may be removed) is that dead loads are known with much more certainty than live loads. This means that a lower factor of safety is acceptable for dead loads than for live loads.

Consider the floor of a lecture room. Let us assume that the maximum occupancy of a 6m by 4m room is 40 students with furniture having a mass of say 1 tonne (=1000 kg). If we assume the average mass of a student is 75 kg then the total movable or "Live" Load on the floor is 40 × 75 + 1000 = 4000 kg. Changing the mass into force we divide by 100 to give a total live load on the floor of 4000/100 = 40 kN. The floor structure would then have to be designed to carry this "live load" as well as its own "dead load".

We could do this exercise whenever we designed a structure but it would be tedious and subject to individual differences in assumptions. For that reason the Australian Standard AS 1170.1 "Dead and Live Loads" specifies what values of live load must be assumed by structural designers. As the value quite obviously depends on the area involved AS 1170.1 specifies the design Live Loads in terms of a *PRESSURE = FORCE PER UNIT AREA*. The units of pressure used in structural work are kN per square metre or kiloPascals (kPa). Note that 1 kN/m^2 = 1kPa and that 1kPa = 1000 Pa (however we rarely use the Pascal (Pa) in building work). In the above example the Live Load pressure due to the students and furniture would be 40/(6*4) = 1.67 kPa. We see below (Table 1.3) that in fact Institutional buildings must be designed for a live load of 3 kPa in AS 1170.1.

NOTE   TOTAL LIVE LOAD(kN) = LIVE LOAD PRESSURE (kPa) × AREA (m²)

Table 1.3   Design Live Loads

OCCUPANCY	LIVE LOAD PRESSURE (kPa)	CONCENTRATED LOAD (kN)
Homes	1.5	1.8 on 350 mm^2
Club Lounges	5.0	3.6
Institutional. Buildings.	3.0	2.7
Print Storage	12.5	9
Offices	3.0	6.7
Shops	5.0	7

When we speak of a "live load" in buildings we are generally talking about Live Load Pressure eg 3 kPa. Remember also that 1 kPa = 1 kN/m^2

In addition to the pressure resulting from a spread of loading on a structure AS 1170.1 requires that structures be designed separately for a concentrated force as given in the code (See Table 2.3). This is to allow for the occurrence of localised heavy loading. Note that these concentrated forces are not additional to the live load pressures but are to be treated as a separate loading case.

Note that AS 1170.1 requires an additional 0.5 kPa of live load where there are movable partitions. It also requires that the design load not be less than the actual live loads if they are known. eg if the live load in a shop is known to be 7 kPa then the structure should be designed for this load, not the 5 kPa given in AS 1170.1.
There is provision in AS 1170.1 for a reduction in live load on certain floors up to a maximum of 50 % (See Clause 4.9)
Live loads on non-trafficable roofs vary according to the area of roof supported by the roof member according to the following formula -

Live Load = (1.8 / A) + 0.12 kPa   where

A = Area of roof supported by member in m^2
Using this formula will give a live load of roughly 0.75 - 1.5 kPa for rafters.

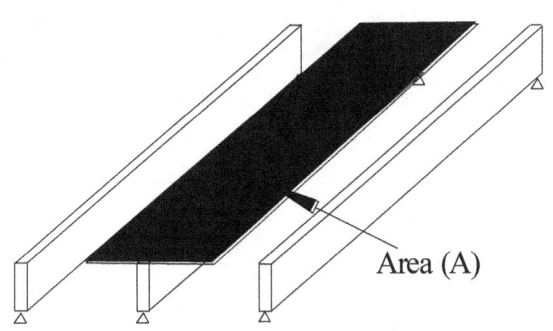

Area (A)

EXAMPLES

1. A floor measuring 4 m by 5m is to be designed for a live load of 3 kPa. What is the total live load on the floor ?

Total Live Load = Area * Live Load Pressure
= (4 × 5) × 3 = 60 kN

2. Rafters are 3 m long and spaced at 450 mm centres. What is the live load per metre run on a rafter (w) and the total live load on the rafter (W)?

Live Load Pressure = 1.8/(3×0.45) + 0.12 = 1.45 kPa

Live Load per metre run (w) = Live Load Pressure×Rafter Spacing
= 1.45 × 0.45 = 0.66 kN/m

Total Live Load (W) = Live Load/metre × Span
=0.66×3 = 1.98 kN

EXERCISES:

1. a)   Calculate the gravitational force of the following reinforced concrete T-beam, assuming it is 5 metres long.

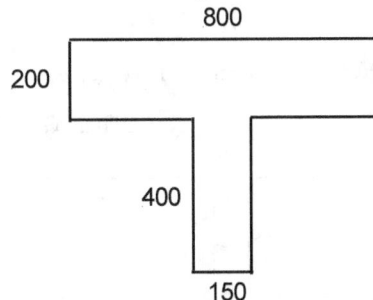

   b)   Calculate the gravitational force of a 4 metre diameter pile of loose sand assuming it slopes at 30° to the horizontal.

   c)   Calculate the gravitational force of a 3m by 4m sheet of 10mm glass.

   d)   Calculate the "average" pressure on the ground due to the pile of soil in b).

   e)   Re-calculate question 1c)using Table 1.2 instead of Table 1.1.

2. Calculate the dead load in kN/m^2 (kPa) of the following floor.

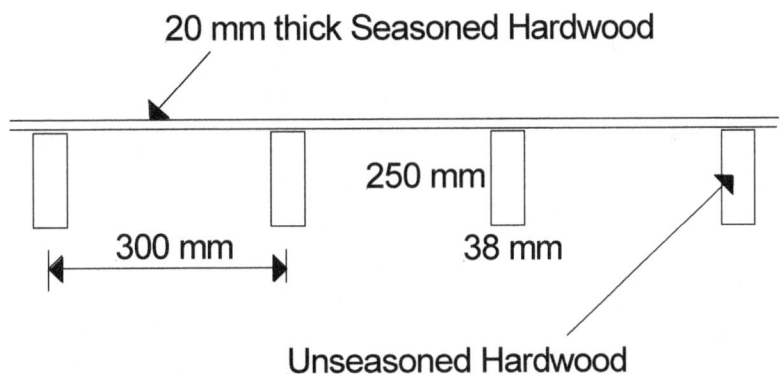

3. Calculate the dead load(Force/m^2) in kN/m^2 (kPa) of the following reinforced concrete floor. Hint: Treat thickenings below slab as you would a joist.

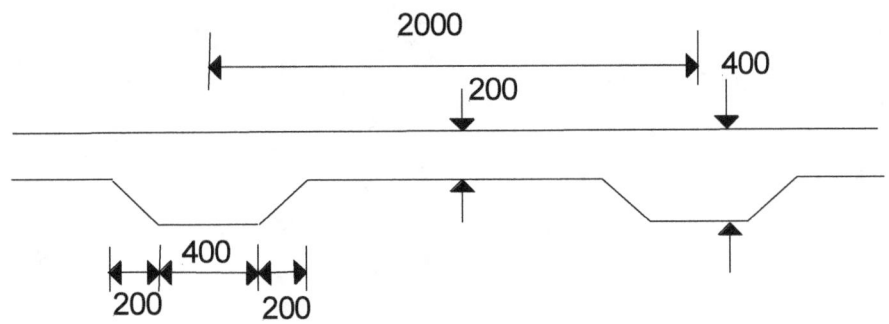

4. What is the total live load in kN on an office floor that is 6m by 4 m?

5. What is the dead load (Force/m^2) and the live load (Force/m^2) of the following roof?

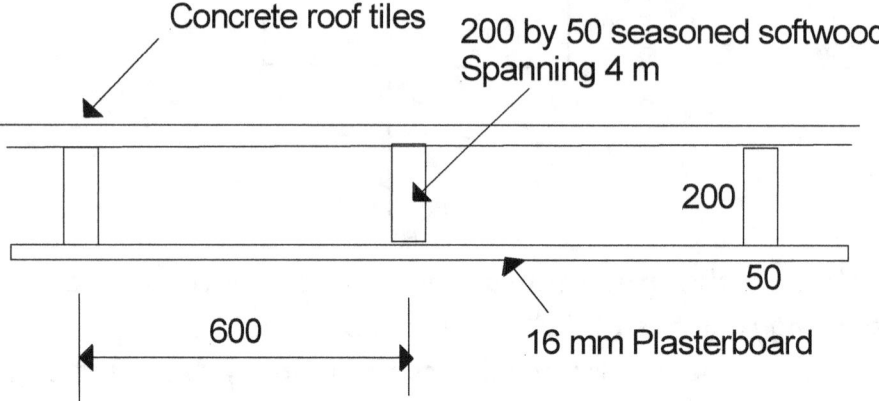

Concrete roof tiles
200 by 50 seasoned softwood Spanning 4 m
200
50
600
16 mm Plasterboard

6. What is the total load (D.L.+L.L.) in kN/m run on the rafters in Q5?

2. OTHER LOADS ON STRUCTURES

2.1 Wind Pressure

For light buildings such as portal frames and tall buildings wind pressure is generally the most important design force. Wind also plays an important part in the overall stability of temporary structures.

Points to note about wind pressures are

1) Design wind pressures vary depending on the location of the building within Australia. For example the design wind pressure for a building in Darwin (Cyclone area) would be roughly twice that for a similar building in Sydney

2) Wind pressure increases in proportion to the wind velocity squared. For example if the wind velocity increases from 20 to 40 m/sec then the wind pressure will increase fourfold $((40/20)^2)$.

3) Wind pressure increases with height, quickly at first and then the rate drops off. The rate depends on the ground roughness but as a general guide in open terrain the wind pressure at 10 m is 25 % greater than at ground level and 70% more at a height of 100 m.

4) Wind pressure drops off with increasing roughness of the terrain. For example a low level building in the inner city would be designed for a wind pressure of approximately 50 % of a similar building in open terrain

5) Design wind pressures vary depending on the shape of the building and can vary from positive pressure to negative pressure (suction). Negative pressures typically occur on flat roofs and on the leeward side of buildings.

6) Design wind pressures can vary depending on the orientation of the building and this can be utilised in design if sufficient data is available.

7) Design wind pressures are based on gust wind speeds. Where structures are temporary the chance of getting that design wind speed is reduced and hence a lower wind pressure can be used in design.

2.1.1 Basic Regional Wind Speeds

Basic gust wind speeds for normal buildings are based on an analysis of wind records to find the gust wind speed that has a probability of being exceeded once in every so many years (return period). For strength calculations (i.e. for calculating whether structures are strong enough) the return period is normally chosen as 500 years for normal buildings with a lifespan of 50 years ("V_u"). For serviceability calculations (i.e. for calculating deflections of structures due to wind) the return period is 20 years ("Vs"). For temporary construction structures a return period of 50 years is appropriate for strength calculations.

AS1170.2 "Wind Actions" lists different wind speeds for different areas (Regions) of Australia and for various return periods. For Sydney, the basic regional wind speed (V) is given as

$$V = 67 - 41 \times R^{-0.1} \text{ where R = Return Period}$$

Eg if R = 500 (strength checks) $V_u = 67 - 41 \times 500^{-0.1} = 45$ m/s

If R = 20 (serviceability/deflection checks) $V_s = 37$ m/s

(For the Gold Coast $V = 106 - 92 \times R^{-0.1}$)

(Note 1m/s is approx 2 knots (1.94 exactly) and approx 4km/hr (3.6 exactly)

2.1.2 Site Wind Speeds

The regional wind speeds are changed into site wind speeds (V^*_u or V^*_s) by multiplying by the following factors.

1. Terrain Category Factor (F_1)
   The design wind velocity depends on the surrounding terrain. Terrain Categories are indicative of the roughness of the ground over which the wind blows. They are
   a. Terrain Category 1 (Very Exposed open terrain)
   b. Category 2 (Open Terrain with well scattered obstructions such as around airfields),
   c. Terrain Category 3 (Terrain with numerous closely spaced obstructions such as houses in inner suburban areas
   d. Terrain Category 4 (Terrain with numerous large and high closely spaced obstructions).

   Up to a height of 10 metres the values of F_1 are 1.12 (Terrain Category 1), 1.00 (Terrain Category 2), 0.83 (Terrain Category 3) and 0.75 (Terrain Category 4).

2. Height Factor (F_2)
   The variation of design velocities <u>over 10 metres</u> high depends on the terrain category. For Terrain Category 2 the following formula can be used
   $$F_2 = 0.86 \times (\text{Height})^{0.077}$$
   $$= 1.0 \text{ for Heights less than 10 m}$$

   For example at 100 metres the height factor is $0.86 \times 100^{0.077} = 1.23$
   The height factor varies slightly for other terrain categories.

3.   Topographic Class (F_3) and Shielding Factors. (F_4)

Structures on hill slopes need to be designed for increased wind velocities and structures flanked by other structures for lesser velocities. Correction factors for these are given in the Australian Wind Loading Code (AS 1170.2). For example F_3 = 1.57 for structures at the top of steep ridges and F_4= 0.8 for fully shielded structures .

$$V^* = F_1 \times F_2 \times F_3 \times F_4 \times V$$

Example: A structure in Sydney is to be designed for strength and serviceability. It is on the flat and is fully shielded by surrounding structures. It is less than 10 metres high and in Terrain Category 1.

$$V^*_u = 1.12 \times 1.0 \times 1.0 \times 0.8 \times 45 = 40.3 \text{ m/sec}$$

Note that in the Wind Load Code for housing discussed in Chapter 10 this example would be classified as Category N2 (See Table 10) with an assumed design wind speed of 40 m/sec.

For serviceability $V^*s = 1.12 \times 1.0 \times 1.0 \times 0.8 \times 37 = 33.2 \text{ m/sec}$

2.1.3 Basic Wind Pressures

Site wind speeds are changed into Basic Wind Pressures by the following calculation

$$\text{Basic Wind Pressure (p)} = 0.0006 \times V^2$$

Eg For the above example

$p_u = 0.0006 \times 40.3^2 = 0.97$ kPa for strength calculations (R=500 years)
$p_s = 0.0006 \times 33.2^2 = 0.66$ kPa for serviceability calculations (R=20 years)

2.1.4   Converting Basic Wind Pressures to Design Pressures

The basic wind pressure must be multiplied by an "Aerodynamic Shape Factor" or "Pressure Coefficient" (C_d) which depends on the shape, porosity and size of the building and from what direction the wind is blowing. For example if we are looking at a signpost with its base on the ground (pressure coefficient of 1.2) located in Sydney in Terrain Category 2 (F_1=1) and with F_2, F_3 and F_4 =1 then p_u = 1.22 kPa (0.0006×45^2) then the design wind pressure for strength would be 1.2×1.22 = 1.47 kPa.

For portal frame type buildings there are various pressure coefficients for each external surface and one for internal pressure (which could be positive or negative depending on the location of openings). To get the design wind pressures on each external element (eg windward rafter) the internal and external wind pressure coefficients need to be added as shown in Figure 2.1 before multiplying by the

13

basic wind pressure. Values on Figure 2.1 are for openings on the windward wall and a roof pitch of 15 degrees.

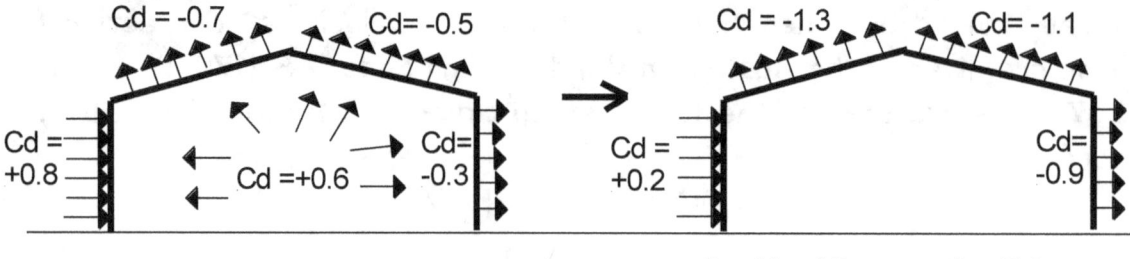

Combined Pressure Coefficients

Figure 2.1 Addition of Wind Pressure Coefficients

Looking at Figure 2.1 we see that on the windward rafter there is an internal C_d of +0.6 (wind pushing out) which is then added to the external C_d of -0.7 (outward suction) to get a resultant C_d of -1.3 for the rafter.

Example:

Consider a portal frame building in Sydney in Terrain Category 3 that is to be checked for lateral movement under wind pressure. It is less than 10 metres high and is located on the flat with no shielding

Then V^*_s =0.83×1×1×1×37 m/s = 31 m/s and p_s = 0.0006×31² =0.58 kPa. Assuming openings on the windward wall and a roof pitch of 15 degrees as in Figure 2.1 the resultant pressures on the surfaces would be.

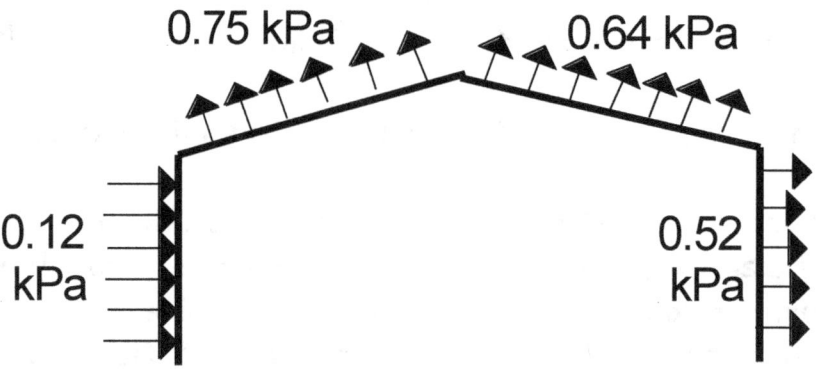

Note 0.75 kPa = 1.3×0.58 etc

14

2.2 Soil Loads

Soil loads are generally expressed as "pressures" with units of kiloPascals (kPa) or (kN/m^2). (Note that Force = Pressure × Area or Pressure = Force per Unit Area).

Soil pressures acting on a structure can be divided into two types

1) Those actively pushing on the structure eg in the case of retaining walls. These are generally lateral pressures but may be vertical pressures as in the case of expansive soils. They may be thought of as "active" pressures.

2) Those reacting to the force of the structure eg the soil pressure underneath a footing is a reaction to the load of the structure above. If there was no structure there would be no pressure. These are generally vertical pressures but may be lateral as in the case of the pressure on a key of a retaining wall which resists sliding. They may be thought of as "passive" pressures.

2.2.1 Retaining Walls

Granular (Sandy) Soils
Lateral pressures behind retaining walls with granular backfill increase with depth. The value of maximum pressure (p_{max}) is given by the following equation,

$$p_{max} \text{ (kPa)} = K_a \times \gamma \times H$$

where $\quad$ K_a = Soil Pressure coefficient
$\quad\quad\quad\quad$ γ = Unit Weight of Soil in kN/m^3
$\quad\quad\quad\quad$ H = Depth below surface in metres

For the purpose of assessing the stability of a wall we can replace the pressure diagram by a force diagram, with the force being equal to the average pressure ($0.5 \times p_{max}$) times the height of the wall, as shown in Figure 2.2.

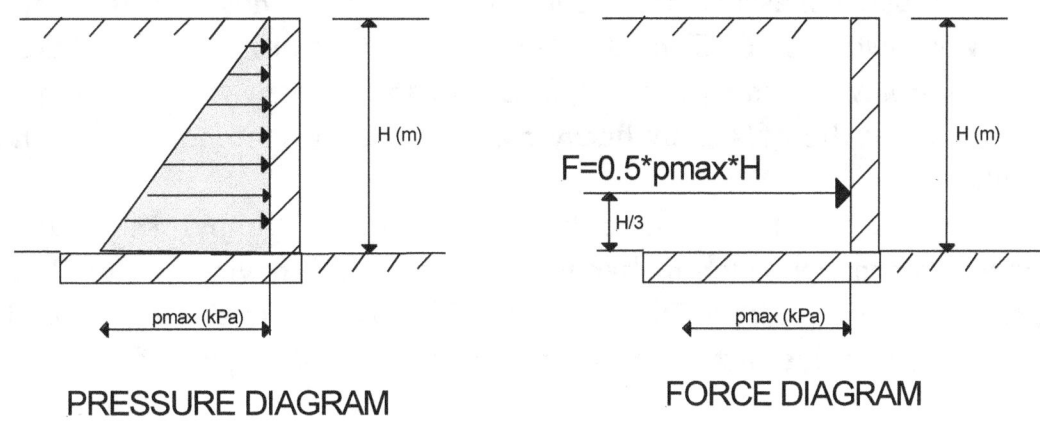

$\quad\quad\quad$ PRESSURE DIAGRAM $\quad\quad\quad\quad\quad$ FORCE DIAGRAM

Figure 2.2 Pressure and Resultant Force on Retaining Walls

For a preliminary analysis the values of K_a and γ shown in Table 2.1 can be assumed for granular soils

Table 2.1   Unit Weights and Soil Pressure Coefficients for Granular Soils

	$\gamma(kN/m^3)$	K_a
Loose sand, gravel,sand & gravel , or ballast	18	0.33
Compact sand, gravel,sand & gravel , or ballast	21	0.27
Broken brick filling	17	0.27

Clayey Soils

Clay soils are not recommended as backfill for retaining walls due to the high pressures generated as clays become wet and also because they do not have the ability to easily get rid of water pressure on the back of a wall. Should it not be possible to use granular backfill then a drainage layer should be constructed immediately behind the wall (commercial plastic "egg-crate" sheets covered in filter cloth are suitable) . In this case K_a may be taken as 1.0 and a unit weight of 20 kN/m^3 assumed.

Notes:

Normal design for retaining walls assumes that the soil is adequately drained. If the soil is not drained an additional water pressure of 10 × depth (kPa) needs to be added to the soil pressure.

The effect of a surcharge eg road traffic acting near the top of the wall is taken care of by increasing the height by a distance equal to the surcharge pressure divided by the unit weight of the soil. For example light road traffic may be considered to cause a pressure of around 3 kPa and this translates to an additional "effective" wall height of 0.15 m (3/20) if the unit weight of soil is taken as 20 kN/m^3. For highway retaining walls it is normal to increase the wall height by 1.2 metres to allow for the effect of heavy concentrated wheel loads acting near the face of walls.

The above formula for p_{max} assumes that the backfill behind the wall is level. Sloping backfills produce much higher pressure eg a wall having a K_a of 0.33 with a level backfill will have a K_a of 0.75 if the backfill slopes upwards away from the wall at an angle of 30 degrees to the horizontal (K_a = 0.40 for slope = 15°).

2.3 Dynamic Loads

Most permanent loads on structures (except wind) are static loads i.e. they do not vary significantly over time. Dynamic loads, on the other hand, are loads which occur due to the interruption of the velocity of a body over a short period of time. Some of the kinetic energy of the body is expended in the form of heat, sound and friction with the remainder being absorbed by the structures concerned.

Dynamic loads occur frequently during construction eg when a crane lifting a load at a certain speed stops suddenly the effect is a dynamic load which is far greater than the static gravitational force of the load.

In dealing with dynamic loads one approach is to resort to the concept of an equivalent static load which is defined as follows

EQUIVALENT STATIC LOAD
= DYNAMIC FACTOR × GRAVITATIONAL FORCE

Consider the case of a 100 kg man (1 kN) standing on a plank, which then deflects an amount, say 10 mm.

If this same man then jumps up and down on the plank, the plank will deflect more than 10 mm say 30 mm. In this case the Equivalent Static Load is 3 kN (30/10 × 10 kN) or the Dynamic Factor is 3. To put it another way, to produce 30 mm deflection using static formula one would have to insert 3 kN into the formula.

Exercise: Suspend a weight on an elastic band. Measure the initial length and the extension when the load is gradually released. Next hold the weight at the un-extended position and release rapidly. You should find that the weight bobs up and down and then settles at the previous extension. The maximum deflection at first "bob" divided by the final extension represents the dynamic factor.

2.4   Earthquake Loads

Earthquake waves produce horizontal and vertical movements of the ground. These movements are typically expressed in terms of the acceleration of the ground as a percentage of the gravitational acceleration "g" (Exact value 9.8 m/s^2). The actual accelerations experienced in a structure depend on

1. The magnitude of the earthquake as measured by the Richter scale. Note each unit on the Richter scale represents an increase in energy of 10, thus an 8 Richter earthquake has 100 times the energy of a 6 Richter earthquake.
2. The ground between the bedrock and the surface at the site. In general soft soils increase the magnitude of the surface horizontal ground acceleration by about 50%
3. The distance of the earthquake epicentre from the site and the type of earthquake waves generated by the earthquake. For this reason ground acceleration at the site is a better measure of the intensity of the earthquake <u>at the site</u>. For example the 2010 Christchurch earthquake (magnitude 7.1 on the Richter scale) resulted in a peak ground acceleration of 0.18 to 0.28 g in Christchurch CBD whilst the much worse 2011 earthquake (magnitude 6.3) produced a peak ground acceleration in the CBD of 0.57 to 0.8g (A maximum of 2.2 g was recorded at Heathcote Valley Primary School).

The peak ground acceleration is represented by the Modified Mercalli Scale shown in Table 2.2. So for example an earthquake with a peak ground acceleration of 0.80 g would be classed as a MM9 earthquake with violent ground shaking.

Table 2.2   Mercalli Earthquake Scale

Mercalli Intensity	Peak Ground Acceleration	Perceived Shaking
1	<0.0017 g	Not Felt
2-3	0.0017-0.014 g	Weak
4	0.014-0.039 g	Light
5	0.039-0.092 g	Moderate
6	0.092-0.18 g	Strong
7	0.18-0.34 g	Very Strong
8	0.34-0.65 g	Severe
9	0.65-1.24 g	Violent
10	>1.24 g	Extreme

4. .The type and stiffness of the structure. Stiff buildings such as masonry buildings are generally subjected to higher accelerations and are therefore more vulnerable.

Figure 2.3   Damage to Masonry Building in 1994 Los Angeles Earthquake

Horizontal acceleration of buildings produce horizontal forces equal to the mass of the building times its horizontal acceleration. Maps in the earthquake code define what accelerations should be assumed for certain areas and the code also gives modification factors for soil conditions and structure stiffness. The magnitude of the design horizontal acceleration on structures is typically 10 – 20 % of "W" but may rise to as high as 0.8 in some circumstances.

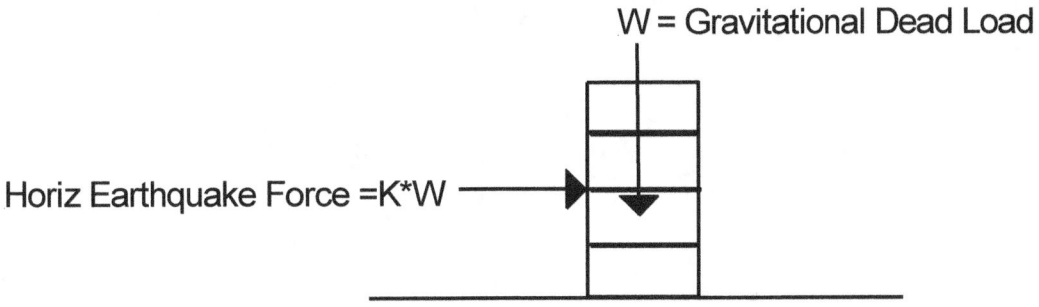

Vertical acceleration of buildings due to earthquakes can be significant and produce increase vertical loads on the footings. This can lead to liquefaction of soft soils and significant settlement leading to failure of buildings.

EXERCISES:

1. Calculate the design wind pressure (for strength calculations) on a permanent hoarding sitting on the ground on the Gold Coast. Assume $C_d=1.2$, Terrain Category 1 and $F_2=F_3=F_4=1$.

2. A 4 metre high retaining wall is backfilled with broken bricks. Calculate the maximum pressure on the wall and draw pressure and force diagrams.

3. For the portal frame shown in Figure 2.1 calculate the wind pressure on the windward wall for strength calculations. Assume structure on the Gold Coast in Terrain Category 4, is fully shielded and located on the top of a hill..

4. Why is the Mercalli scale a better indicator of earthquake damage than the Richter scale?

3. SITE INVESTIGATION AND DESIGN OF SIMPLE FOOTINGS

Figure 3.1 The Leaning Tower of Pisa

The famous Leaning Tower of Pisa was built on unstable ground which caused one side to settle more than the other and this settlement got progressively worse until it posed a danger to tourists. It was stabilised late last century by removing soil from underneath the higher side.

3.1   Site Investigation Techniques

Purpose of Soil Exploration

1. To determine what type of foundation is required
2. To determine excavation difficulty
3. To determine soil properties
4. To establish ground water level

Types of soil exploration include
1.   Visual examination of soils in cuttings
2.   Excavation of test pit.
3.   Hand auger
4.   Continuous hollow spiral flight auger

Most commercial soil investigations are conducted using truck mounted continuous hollow spiral flight augers. In this test a continuous stream of soil is brought to the surface and the colour and type of soil is logged.

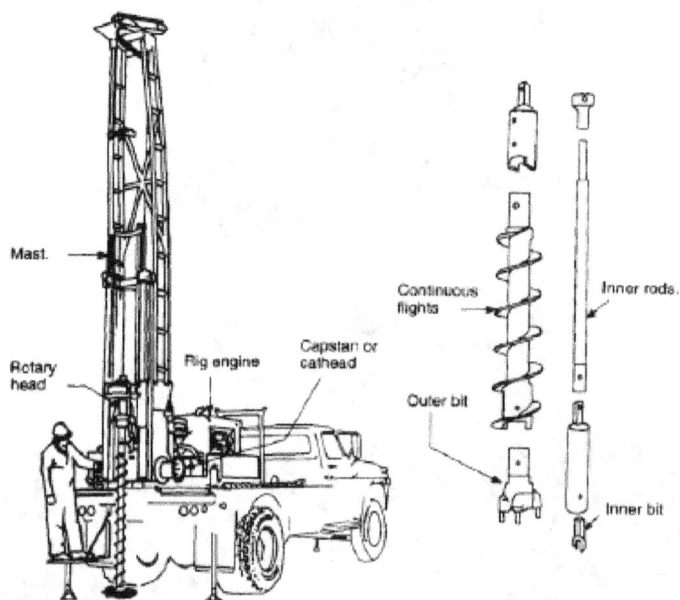

Figure 3.2   Continuous Spiral flight Auger (www.denichsoiltest.com)

Samples taken may either be "disturbed" or "undisturbed". Disturbed samples are obtained by withdrawing the auger and hammering down a split tube into the soil. The test is called the SPT test and the number of blows to hammer down the tube 300 mm is termed the "N' value. and this gives a reasonable guide to soil properties. Where undisturbed samples are required a thinner walled tube (Shelby tube) is used and the split sampler is pushed into the soil rather than hammered.

3.2   Classification of Soils

Soils are usually classified according to their sand, silt and clay contents. To get a quick idea of the various components you can put a sample of soil in a jar with some water. Shake it up and three layers corresponding to the three component types, sand, silt and clay should be able to be identified. Note that the clay particles may take a long time to settle.

Settling Times   Sand 30-60 seconds
                    Silt   15-60 minutes
                    Clay 2 hours –2 days

A more detailed analysis is carried out using a set of standard sieves, with the clay and silt components being identified using a hydrometer. The results are plotted up on a logarithmic chart showing grain size and percentage of sample passing that grain size.

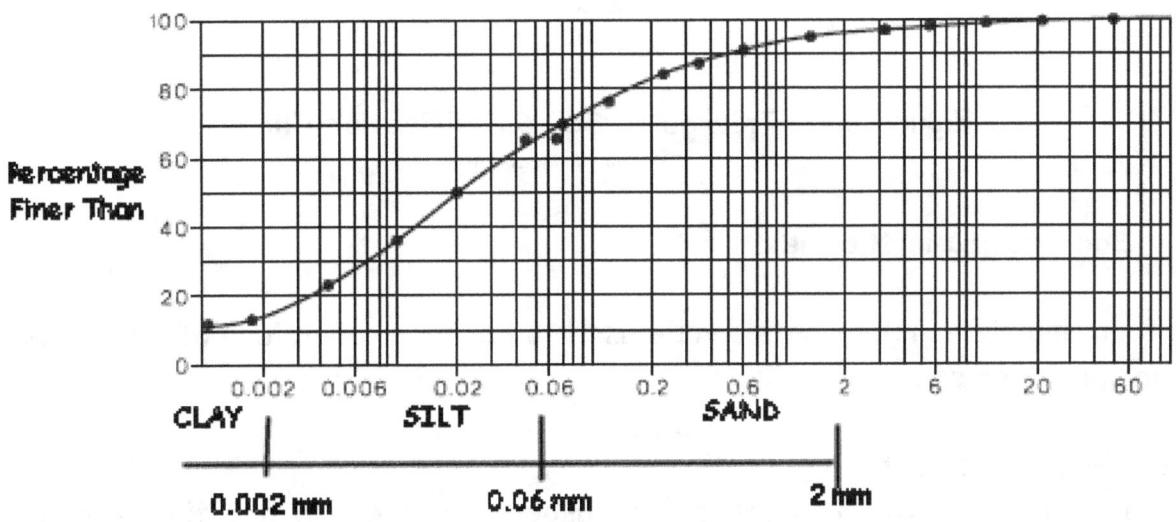

Figure 3.3   Particle Size Distribution

For example in Figure 3.3 above roughly 70% of the sample is smaller than sand size (Meaning 30% sand plus a bit of gravel) and about 15% smaller than silt size (meaning 15% clay and hence 70-15 = 55% silt). This would identify the soil as a "Silty Loam" with the aid of the "Triangular Soil Classification Chart" shown below.

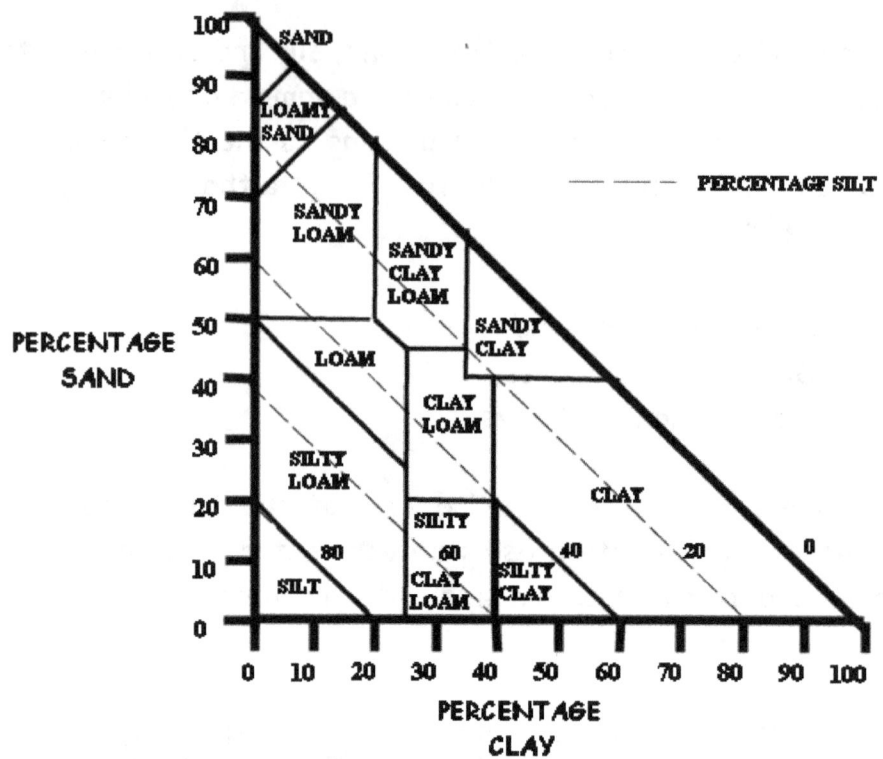

Figure 3.4 Triangular Soil Classification Chart

(©Kevan Heathcote)

3.3   Soil Testing in the Field

The following field instruments can be used to obtain a rough idea of the allowable bearing pressure of soils.

3.3.1 Shear Vane Test

A shear vane as shown in Figure 3.5 is pushed into the clay and the reading for undrained shear strength (c_u) taken when the vane "shears" through the clay when the top is rotated.

Figure 3.5 Shear Vane

3.3.2   Perth (Scala) Penetrometer

In this test the tubular weight is lifted and allowed to drop onto the circular stop, thus forcing the tip into the soil. The number of blows (drops) per 150 mm penetration is recorded for the various depths and is an indication of the density of the soil.

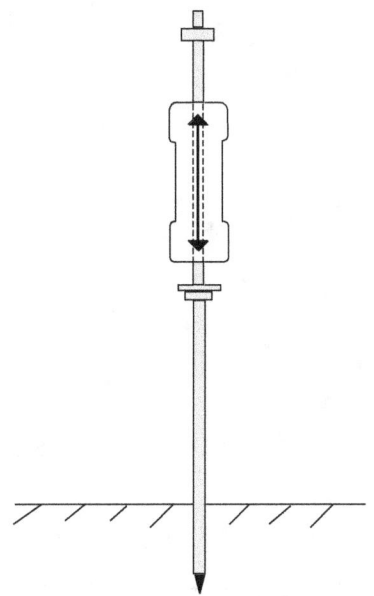

Figure 3.6   Perth Penetrometer

3.3.3   Spiral Flight Auger and SPT Test

This is the standard method used in soil reports. A spiral flight auger is drilled into the ground and the extruded soil is used to provide a description of the various soil layers. Figure 3.7 is a sample bore log reporting the results of a spiral flight augur test hole.

At chosen intervals a "SPT" test is carried out, involving driving a tube into the soil and counting the number of blows of the hammer to advance the tube three successive intervals of 150 mm.

The SPT test is similar to the Perth penetrometer but is a more sophisticated test. Each SPT value represents the number of blows per 150 mm and the "N" number is the sum of the last two readings eg readings of 4,9 ,12 give an "N" of 9+12 = 21.

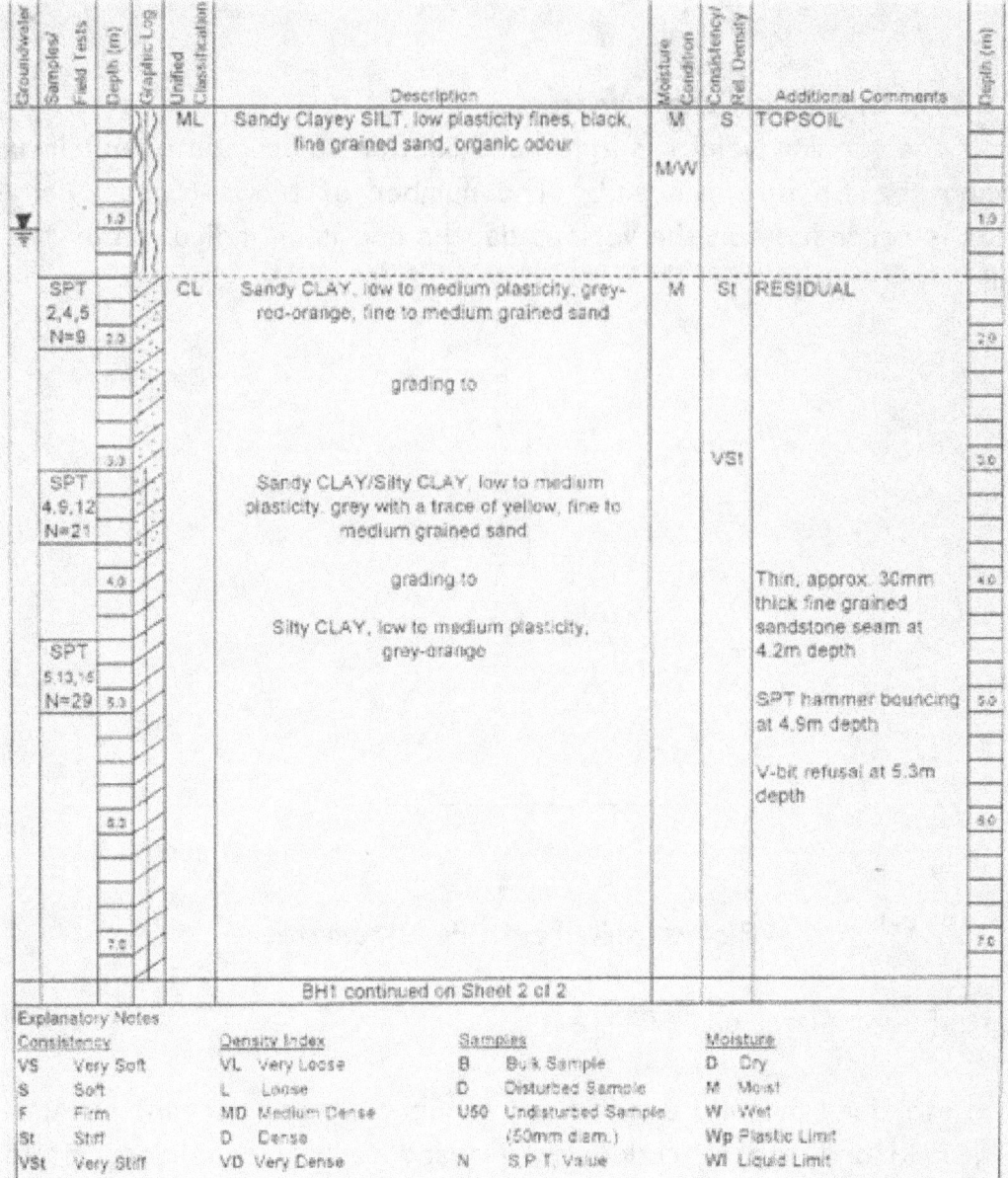

Figure 3.7 Sample Bore Log

Points to note about this bore log (Figure 3.7) are

a) the depth to the water table (about 1 metre)
b) descriptions of soil layers and their extent eg sandy clayey silt from 0 to 1.4 metres
c) consistency of soils eg above soil layer "soft"
d) SPT readings .
e) Depth of V-bit refusal. This represents the depth at which the spiral flight auger cannot penetrate the rock below with a V-bit.

3.4 Allowable Bearing Pressures (ABP)

The allowable bearing pressure on the soil under a footing is termed "Allowable Bearing Pressure" or ABP. This is equivalent to the maximum load in kN that can be carried by the foundation for every square metre of area contacting the soil.

The soil beneath footings can fail in two different ways

1. In <u>shear</u> where the soil particles slide over one another producing a shear failure plane.

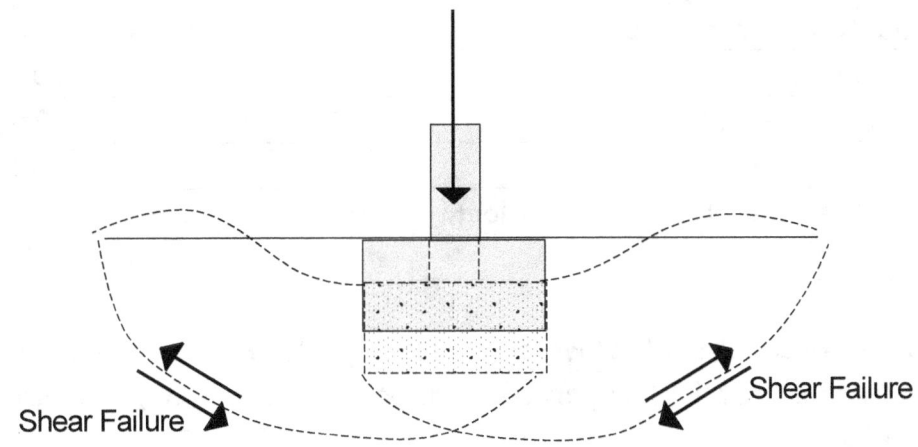

Figure 3.8   Soil Failure Under Footing

In sandy soils the pressure required to fail the soil increases with the footing width and depth below ground level. In clay soil it is purely dependent on the cohesion of the clay. The allowable pressure on the soil is equal to the failure pressure divided by a factor of safety, usually 2 for sandy soils and 3 for clays.

2. In <u>deflection (settlement)</u> where the soil beneath the footing compresses. Sandy soils have a permanent instantaneous settlement whilst clayey soils have an instantaneous settlement which increases over time to a maximum value.

The "Allowable Bearing Pressure" or ABP is taken as the lower of the above two values, usually the settlement criteria. The allowable capacity of a soil based on a settlement criteria is dependent on the allowable settlement, usually taken as 25 mm.

For preliminary design purposes the values for allowable bearing pressure given in Table 3.1 may be used.

Table 3.1 Approximate Values for Allowable Bearing Pressures

Material	ABP (kPa)
Weak sandstone or shale (V bit refusal)	600
Compact well-graded gravel/sand mixtures	400
Very stiff clays and hard shaley clays	400
Stiff clays, sandy clays or sandy clay loam	225
Loose well-graded gravel/sand mixtures	200
Compacted coarse sands	200
Firm sandy loam	150
Loose coarse sand	125
Firm sandy clays	100
Loose fine sand or loose sandy loam	75
Soft Clays	50

For slightly more accurate values we can correlate the Perth Penetrometer results to the ABP. In this test 1 Blow per 150 mm penetration is approximately equal to an ABP of 25 kPa.

More accurate results are obtained by correlating the SPT "N" value to ABP. In this test the ABP is approximately $15 \times N$ for sandy and sandy clay soils.

For very clayey soils the shear vane test gives an ABP approximately equal to 4 times the undrained shear strength (c_u) recorded on the dial.

3.5   Loads on Foundations

The loads to be transferred from the footings to the foundation material include the dead load of the structure, any live loads to be considered and the dead load of the footings.

For simple residential structures the design loads given in Table 3.2 may be used as a guide for <u>full brick</u> construction (two skins external and single internal). For brick veneer construction the loads can be reduced by 6 kN/m for single storey and 12 kN/m for double storey. The given loads include an allowance for the dead load of the footing.

Table 3.2 Approximate Loads on Residential Footings

Building Type	External Walls		Internal Walls	
	Footing Load	Offset	Footing Load	Offset
Single storey house with suspended timber floor	26 kN/m	50 mm	22 kN/m	25 mm
Single storey house with suspended concrete floor	29 kN/m	50 mm	28 kN/m	25 mm
Double storey house with suspended timber floors	44 kN/m	50 mm	36 kN/m	25 mm
Double storey house with suspended concrete floors	54 kN/m	50 mm	50 kN/m	25 mm

3.6 Design of Simple Footings

Footings for a structure can be simple pad footings, strip footings, integrated strip footings eg (slab on ground) or pile footings. Footings for retaining walls are a variation of strip footings and may in some cases be integrated with an adjoining slab.

3.6.1 Strip and Pad Footings

Strip and pad footings are designed so that the calculated bearing pressure due to the applied loads is less than the ABP. Note that when checking bearing pressures, allowable stress/pressure design is used and loads are not factored up as they are for checking out the strength of other members in a structure.

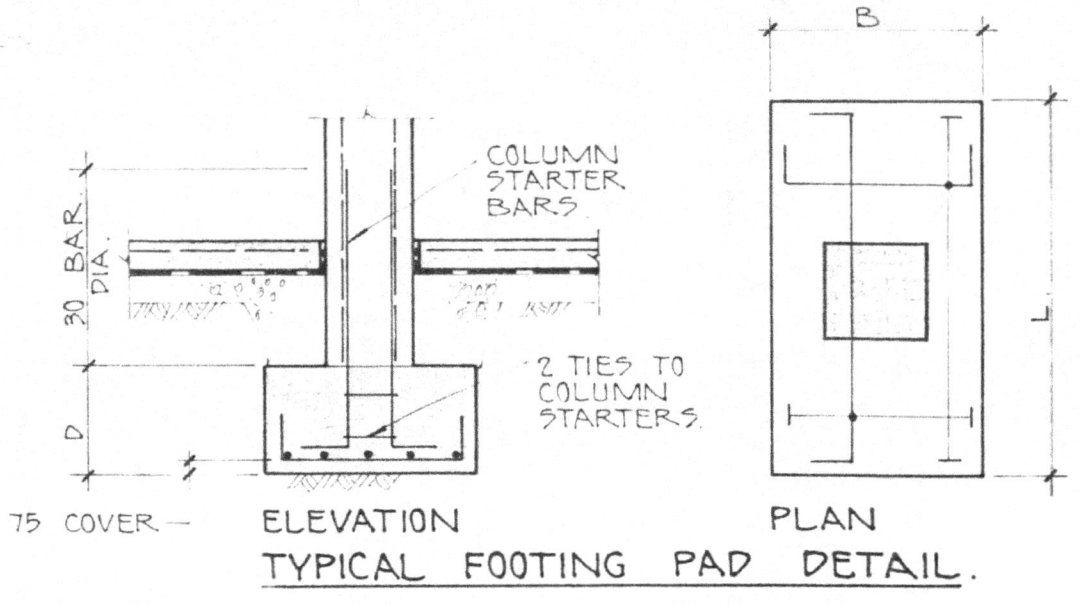

ELEVATION   PLAN
TYPICAL FOOTING PAD DETAIL.

Figure 3.9 Typical Pad Footing Detail

29

Consider the case of a 600mm by 600 mm pad footing carrying a load of 70 kN onto a soil which has an ABP of 200kPa.

Then Actual Bearing Pressure = 70/(0.6×06) = 195 kPa and since this is less than the ABP of 200 kPa the footing is safe. Looking at it another way the safe load of this footing is 200×(0.6×0.6) = 72 kN.

Where footing loads are eccentric the bearing pressure above must be calculated using an 'effective width' (B_e) to take into account the higher pressure that will occur at the edge closest to the applied load.

For granular(sandy) soils $B_e = B \times [1-2 \times (e/B)]$

For clayey and sandy clay soils $B_e = B \times [1-4 \times (e/B) + 4 \times (e/B)^2]$

where e is the distance from the centre of the footing to the load and B is the footing width.

So in the above case if the load of 70 kN was 100 mm eccentric (e/B = 100/600 =0.167) then if the soils was granular B_e = 600×[1-2×0.167] = 400 mm.

In this case the calculated maximum pressure would then be 70/(0.4×0.6) = 292 kPa and since this is greater than the ABP the footing is unsafe. Again expressed another way the safe load of this footing would be 200×(0.4×0.6) = 48 kN

Situations like this commonly occur when footings are on a boundary and "strap' footings (See Figure 3.10) are used linking the boundary footing to other footings further from the boundary to even out the pressure.

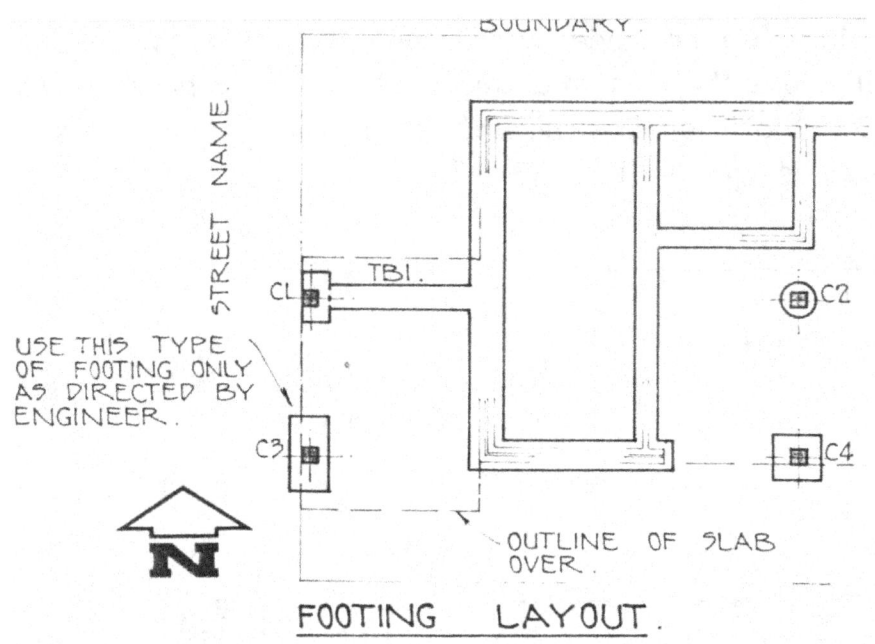

FOOTING LAYOUT.

	1300 K.Pa INTERNAL 860 K.Pa BOUNDARY					COLUMN SIZE	COLUMN STARTER BARS
COLUMN NUMBER	FOOTING SIZE			REINFORCEMENT.			
	L	B	D	LONG BARS	SHORT BARS		
C1	1800	1000	900	5C16	2C16	350×350	4C20
C2	Pier Type			P1.		350×350	4C20
C3	2500	1200	1050	7C16	14C16	350×350	8C24
C4	1500	1500	1000	8C16	8C16	350×350	8C24

FOOTING SCHEDULE

Figure 3.10 Footing Details

3.6.2 Pier Footings

Auger bored piers are commonly used on building sites where suitable foundation material is not close to the surface. Such piers are commonly carried to rock where the capacity of the pier is determined by the ABP of the rock. In some circumstances piers are "belled" out to a larger diameter at the rock interface to increase the bearing area and thus the capacity of the pier.

3.6.3 Piled Footings

Piled footings are typically used when there is no rock present. They may be timber, precast concrete or steel and rely on the friction between the sides of the pile and the soil to carry the load. Piles are driven until the number of blows to advance (set) equals the designed value.

3.6.4 Footings for Retaining Walls

Retaining walls are designed for active earth pressures (See Chapter 2). Footings have to be such that the maximum bearing pressure is less than the ABP.

An approximate check can be made by assuming a vertical load (W) equal to the weight of the soil above the footing acting at an eccentricity "e" as was done for eccentrically loaded footings. The value of "e" is dependant on the soil horizontal force (H) and W and is equal to $(H \times L)/(3 \times W)$.

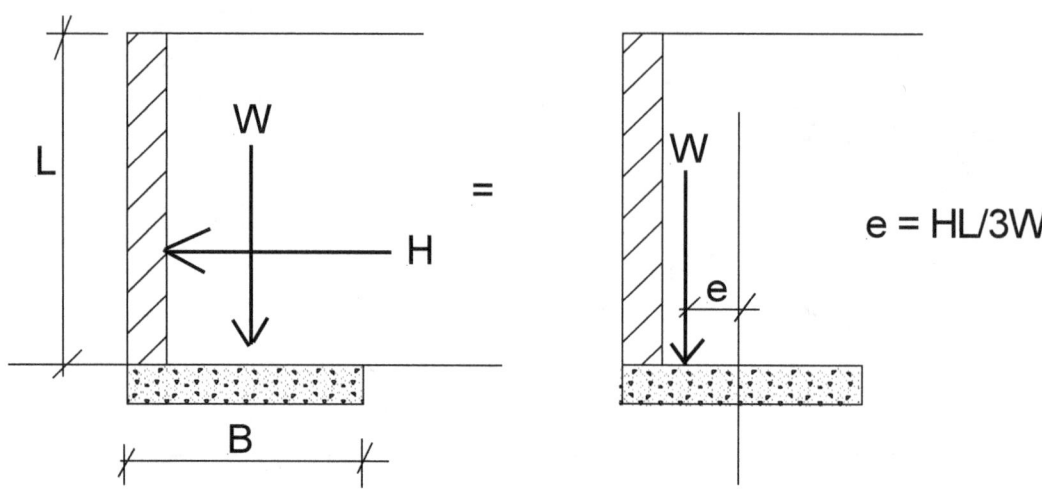

Consider the case of a 3 metre high retaining wall with loose sand backfill ($\emptyset$ =33° and φ =18kN/m^3) on a sandy clay base (ABP =200kPa) with a 1500 mm wide footing.

Then $H = 0.5 \times p_{max} \times$ Height $= 0.5 \times (0.33 \times 18 \times 3) \times 3 = 26.73$ kN
$W = (1.5 \times 3) \times 18 = 81$ kN
and $e = (26.73 \times 3)/(3 \times 81) = 0.33$ m
$e/B = 0.33/1.5 = 0.22$

Then $B_{eff} = 1.5 \times (1 - 4 \times 0.22 + 4 \times 0.22^2) = 0.47$ m

Actual maximum bearing pressure is then $W/(B_{eff} \times 1) = 81/0.47 = 172$ kPa. Since this is less than 200kPa the footing is adequate,.

Of particular concern to building surveyors and builders is the fact that the assumption of active pressure assumes that the wall will rotate when backfilled to a possible top deflection of H/500 for a granular fill. For example a 3 metre high wall may deflect 6 mm at the top. If the backfill was clay the deflection at the top could be as much as 60 mm, which is a good reason not to use clay backfill. Another reason why granular backfill is specified is that it does not retain water whereas the water in a clay backfill imposes an additional load on the wall.

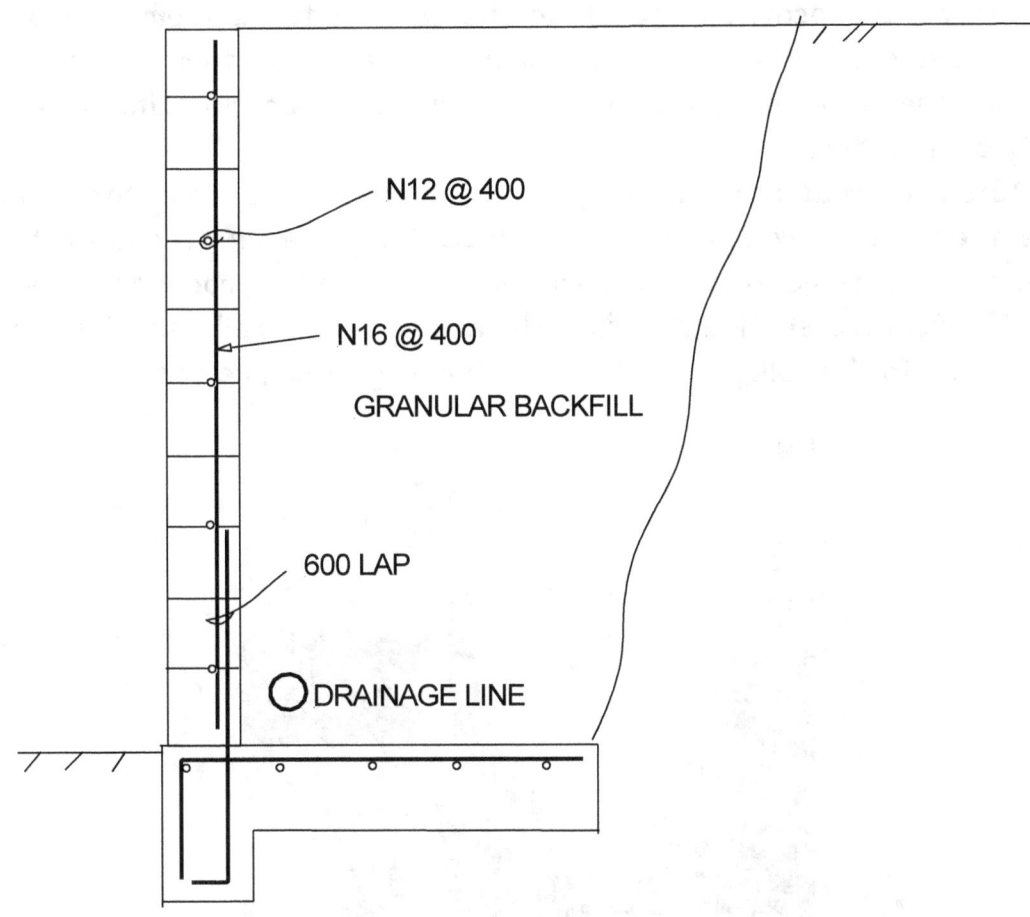

Figure 3.11 Typical Blockwork Retaining Wall

3.7   Footings Too Close To Excavation

In built up areas it is common for excavations for new buildings to extend right up to the boundary and this often causes settlement of adjacent buildings and subsequent legal action.   This problem is avoided if adjacent excavations are located at a safe distance from the zone of influence of adjacent footings (See Figure 3.12).

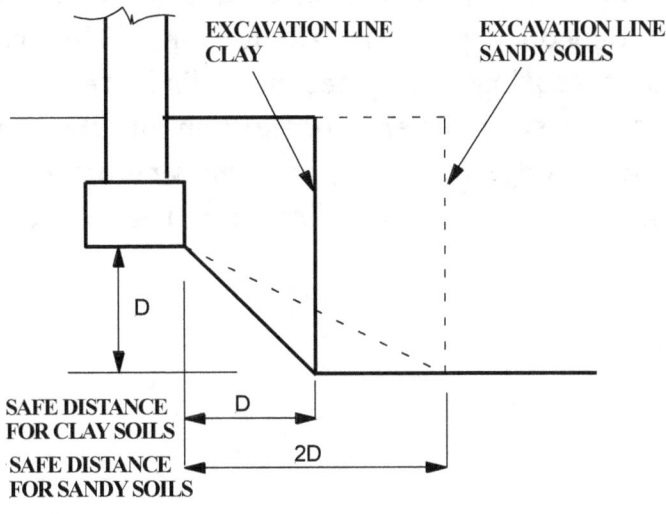

Figure 3.12 Safe Footing Distances from Excavation

Where footings are located closer than the safe distance from the excavation there is a resultant pressure on the side of the excavation. If there is no retaining wall then collapse of the excavation face may occur leading to settlement of the adjacent footings.

Figure 3.13 shows what can happen when you excavate too close to footings. In this case there was a row of terrace houses. The owner of one of them was building a cellar underneath and excavation for the cellar proceeded close to the footings. The footings settled significantly leading to the collapse of the adjacent walls and finally to the collapse of the whole building. Imagine coming home to this!

Figure 3.13 Result of Excavation too Close to Internal Walls

Note that in some cases sheet pile retaining walls (or contiguous concrete piles) are installed on the boundary prior to excavation. This reduces the risk of settlement of adjacent footings but does not eliminate it. As the excavation proceeds there is usually some lateral deflection of the wall with consequent settlement of soil beneath adjacent footings. One way of reducing movement to a minimum is by pouring the ground floor first and then excavating the soil from under the slab.

3.8   Footings on Expansive Soils

Residential footings founded on expansive clay were a major source of building damage prior to the introduction of the Australian code AS 2870 Residential Slabs and Footings in 1988. Prior to this strip footings were generally 300-600 mm wide and 300mm deep and raft slabs (Slab on Ground) footings had edge beams that were also about 300 mm deep. The stiffness of these footings was inadequate to cater for seasonal moisture movements of expansive clays and hence many buildings suffered cracking in their walls. AS 2870 changed the focus from bearing capacity to stiffness of footing and for that reason most footings designed to that code for expansive clay soils are much deeper, ranging from 400 to 1000mm in some cases.

Classification of soils as expansive depends on their tendency to expand when moisture is added or to shrink when moisture is removed. This ability of a clay soil to expand can be approximately related to its "Liquid Limit" which is determined by the Casagrande apparatus. In this test a soil sample of a known moisture content is placed in a small dish and levelled out. A groove is then made through the middle with a special grooving tool. The dish is mounted on a stand which has a handle which enables the dish to be lifted up and dropped onto a rubber pad. This is done 25 times. The test is repeated at different moisture contents until the gap closes at 25 blows. This moisture content is then referred to as the "Liquid Limit" of the soil.
The Liquid Limit is an indication of the moisture content at which the soil changes from a solid to a liquid. Soils which have a low reactivity (i.e. are not expansive) generally have a liquid limit below 35% whilst those with a high reactivity have liquid limits greater than 50 %. There are other tests to more accurately determine the expansion potential of soils and in general the advice of a geotechnical engineer should be sought to classify the soil.

AS 2870 classifies soils as follows

CLASS	FOUNDATION
A	Sand Sites
S	Slightly reactive clay sites
M	Moderately reactive clay sites
H	Highly reactive clay sites
E	Extremely reactive clay sites

In general strip footings are suitable for Classes A and S but for all other classes stiffened raft footings are generally used nowadays with standard designs given in AS 2870.

EXERCISES:

1. A soil has 50% sand and 40% clay. It is firm. What is it called and what would be a reasonable ABP for it ?

2. A 400mm square footing is sitting on the above soil at an eccentricity of 50 mm. What is the maximum load in kN that can be applied to it ?

3. A double storey house has suspended timber floors. The strip footings for the external walls are 400 wide by 400 deep. Is the bearing pressure on the sandy soil OK if the ABP is 150 kPa ?

4. Using the information given on the bore log below determine at what depth you would place a 600mm by 600 mm pad footing if it carries a load of 40 KN which is 100 mm eccentric. (Hint: Calculate the load capacity of the footing at each SPT test position)

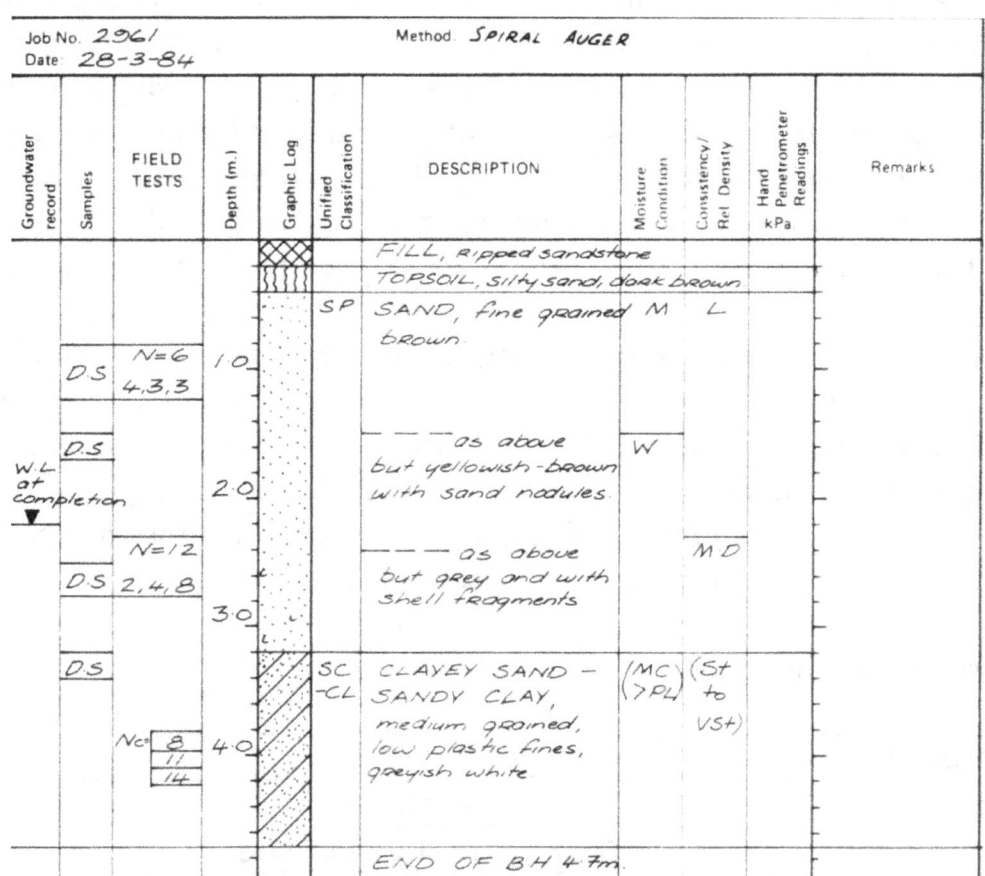

5. A retaining wall is 4 metres high with a base of 2.8 metres. Backfill is compact sand and the base material is stiff clay. Determine the maximum bearing pressure on the clay and state whether it is less than the ABP.

4. STRUCTURAL RESPONSE TO LOADS

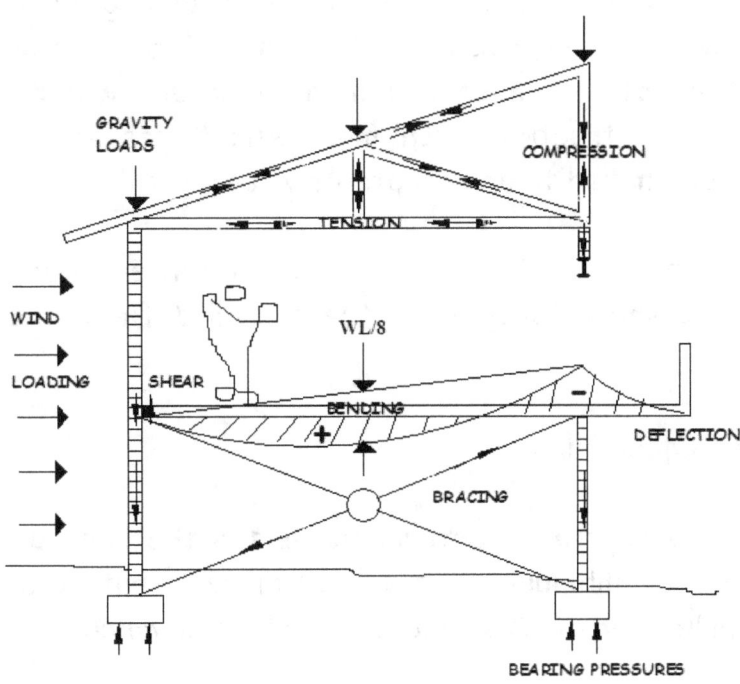

4.1 Introduction

In order to determine whether a structure is safe or not it is necessary to first determine the loads acting on the structure (Chapters 1 & 2) and then determine what effect these loads will have on the structure. External loads on a structure produce various internal effects. They may produce

1. Compression or tension forces within a member eg. Pulling an elastic band will produce a tension force in the band. Structural elements where the only internal effects are either tension or compression are referred to as "trusses" if they are two-dimensional and "space frames" if they are three-dimensional. Tension and compression forces are collectively referred to as "axial forces" as they involve forces acting in the axial direction of the member.

2. Bending of the member eg. If a weight is placed on the end of a cantilever beam it will bend, with the amount of bending being the greatest nearest the support. The extent to which a member will be bent due to the applied forces can be calculated and expressed as a "bending moment". The higher the value of "bending moment" in a member the greater the amount of bending. A graph showing the variation of bending moment along the length of a member is referred to as a "bending moment diagram".

3. A "shearing" action i.e. a tendency for one section of the member to separate vertically (shear) relative to its adjacent section. The magnitude of the shearing force within a member is calculable and can vary along its length. A

37

graph showing the variation of shear force along the length of a member is referred to as a "shear force diagram".

4. Deformation of individual members of the structure eg. Pulling an elastic band will result in the band increasing in length. The combined effect of the deformation of all of the members in a structure will result in an overall "deflection" of the structure. In most structures the value of maximum deflection must be limited to an acceptable value.

This chapter will examine how the magnitude of the internal actions (axial forces, bending moments and shear forces) and the overall deflection of a structure can be calculated.

4.2 Equations of Equilibrium

The most powerful concept in structural analysis is the concept that a stationary structure must be in equilibrium, ie. it will not move or rotate under the action of applied forces. Application of this concept enables the calculation of

1. The reactions of a structure. The reactions are in essence the actions that the structure imposes on the supporting medium eg ground or wall. Such actions are typically forces but may be bending moments as in the case of a cantilever beam where the action on the support wall is one of "bending".

2. The internal actions (bending and shear) within members.

The concept of equilibrium is defined by the following three equations

1. the sum of all the horizontal forces acting on a structural element must equal zero ie. the structural element must not move sideways.
2. the sum of all the vertical forces acting on a structural element must equal zero ie. the structural element must not move vertically.
3. the sum of all the moments about any point on the structural element must be zero i.e the structural element must not rotate.

It is convenient to think about moments in terms of spanners and a nut. The amount of (twisting) moment (M) we apply to a nut using a spanner is a function of the applied force (P) and the length(lever arm) of the spanner (L). In equation form

$$M = P \times L \quad kN.m \ (if \ P \ in \ kN \ and \ L \ in \ metres)$$

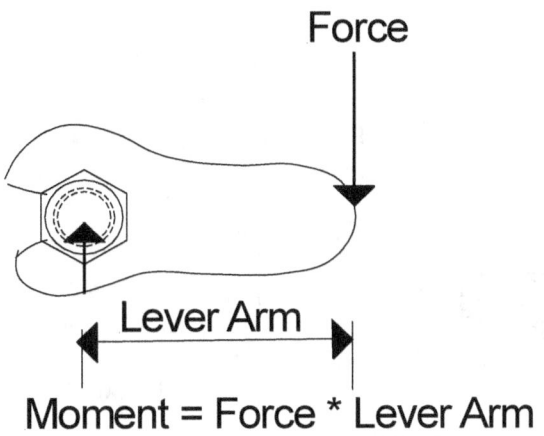

Force

Lever Arm

Moment = Force * Lever Arm

To increase the moment we can either increase P or L. If two spanners are attached to the same nut and one is turned clockwise with force P_1 and length L_1 with the other turned anti-clockwise with force P_2 and length L_2 then if the nut is not to turn, i.e. if it is to be in moment equilibrium then $P_1 \times L_1$ must equal $P_2 \times L_2$.

4.2.1 Equilibrium of Stone Cantilever Steps

Stone cantilever steps do not have stringers and at first glance it is hard to see how they are supported as they are only notched a small distance into the wall, insufficient to provide the necessary cantilever moment capacity

Figure 4.1   Cantilever Stone Stairway (http://www.stone-mason.co.uk)

The diagram below shows how each tread is in equilibrium, with only a torsional moment restraint required at the wall. The weight of each step is progressively carried by the step below and the out of balance moment is resisted by the embedment into the wall. This moment is greatest at the bottom of the stairs

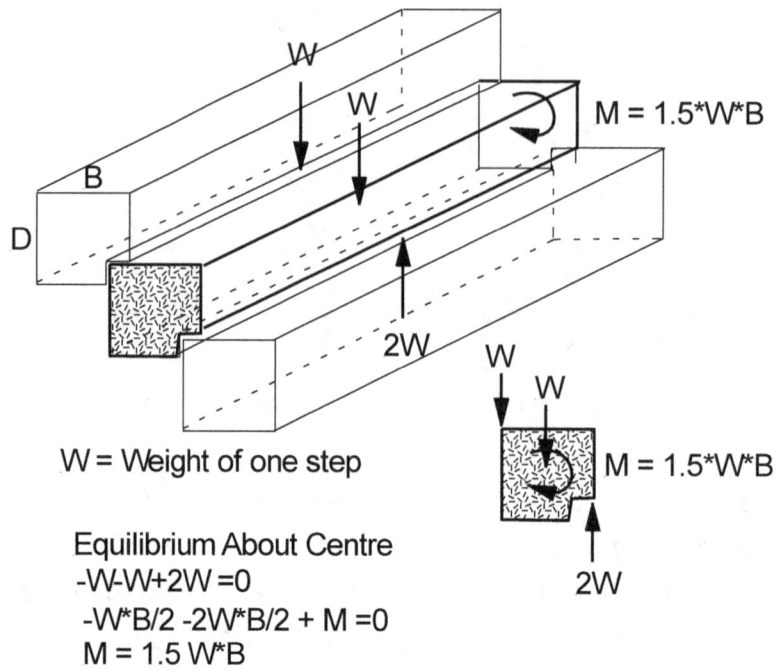

W = Weight of one step

M = 1.5*W*B

Equilibrium About Centre
-W-W+2W =0
-W*B/2 -2W*B/2 + M =0
M = 1.5 W*B

4.3 Reactions of Beams, Frames and Trusses

The above three equations of equilibrium can be used to calculate the reactions of what are called "statically determinate " structures. Statically determinate structures have three unknown reactions at their supports, typically one vertical reaction at each of two supports and one horizontal reaction at one of the supports.   Note that one of the unknown reactions may be a moment as in the case of cantilevers.   Below are some common structures showing where the three unknown reactions are located

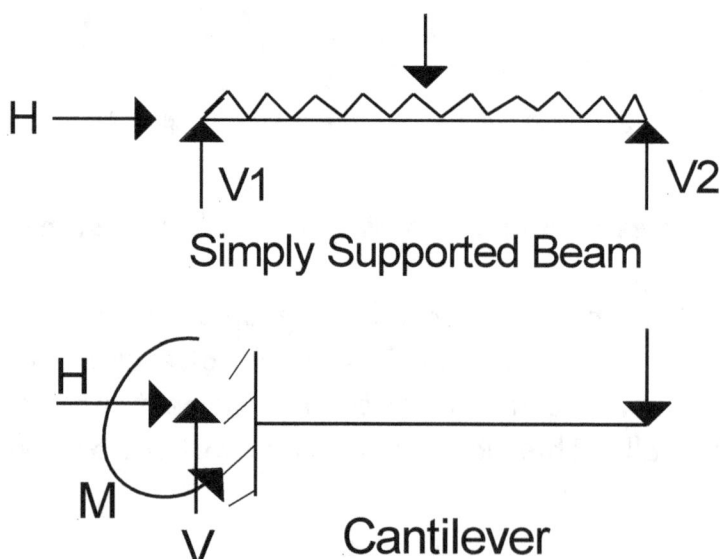

Simply Supported Beam

Cantilever

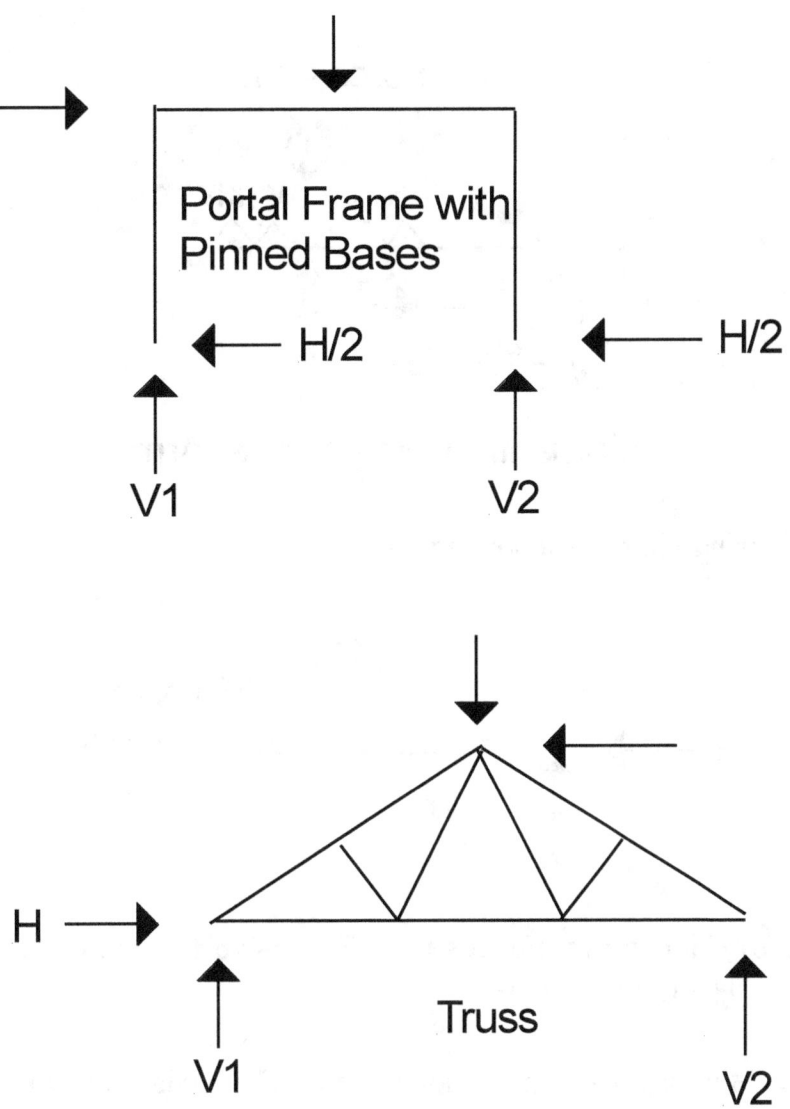

Portal Frame with Pinned Bases

Truss

4.3.1 Reactions of Simply Supported Beams

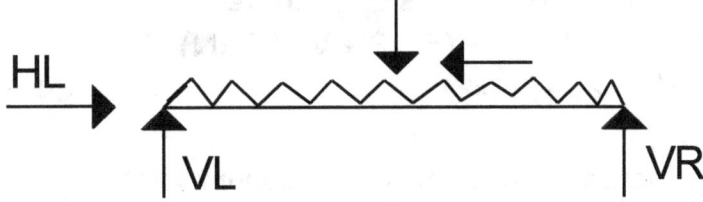

The steps involved in calculating the reactions of a simply supported beam are
1. Sum the horizontal forces to get H_L
2. Take moments about the left hand side to get V_R
3. Sum the vertical forces to get V_L
4. Check that the sum of the moments about the right hand support is zero.

When calculating moments caused by uniformly distributed loads (UDL units kN/m) it is convenient to think of the lever arm of the spanner as being measured from the centre of the load with the force being equal to the UDL times the length of the load as follows.

41

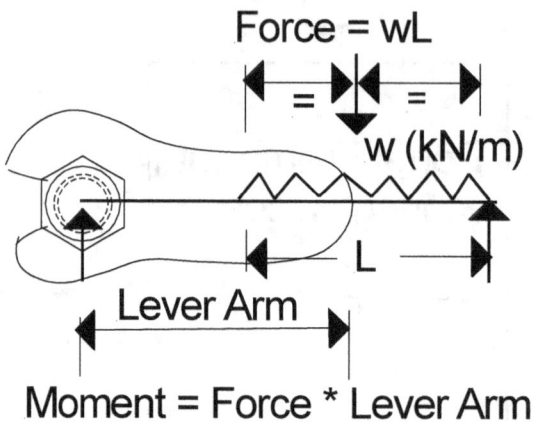

Force = wL

w (kN/m)

L

Lever Arm

Moment = Force * Lever Arm

Consider the following simply supported beam.

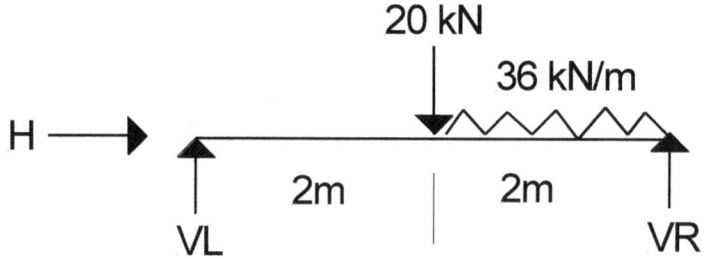

20 kN

36 kN/m

H →

2m

2m

VL

VR

Step 1: Σ (Sum of) Horizontal Forces = 0   Forces to the right positive

H + 0 = 0   » H=0

Step 2: Σ Moments about left hand support = 0   Clockwise moments positive

$(20 \times 2) + (36 \times 2) \times 3 - V_R \times 4 = 0$

$256 - V_R \times 4 = 0$ » $V_R = 256/4 = 64$ kN

Step 3: Σ Vertical Forces = 0   Forces up positive

$V_L - 20 - (36 \times 2) + V_R(64 \text{ kN}) = 0$

$VL = 20 + (36 \times 2) - 64 = +0$ » $V_L = 28$ kN

Step 4: Check Σ Moments about right hand support = 0

$28 \times 4 - 20 \times 2 - (36 \times 2) \times 1 = 0$

So results right.

Now consider the following cantilever beam

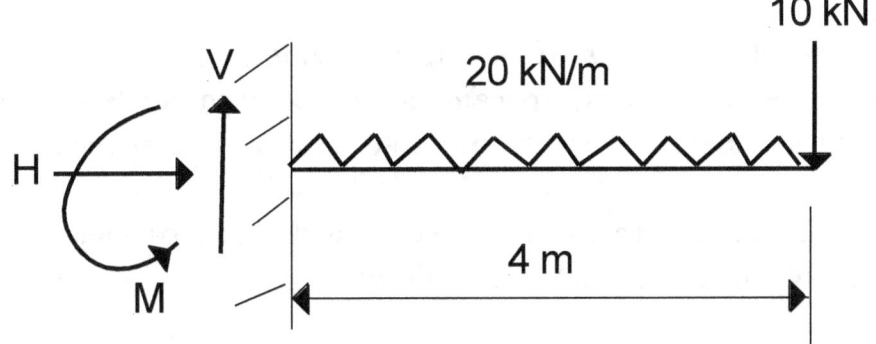

The three unknown reactions are M, H and V

Step 1:

Sum the horizontal forces to get H, forces to the right are positive.

H + 0 = 0

H = 0

Step 2:

Take moments about the left hand side to get M, clockwise moments are positive.

- M +(4×20)×2 + (10×4) = 0

M = 200 kN.m

Note that we could check our answer by taking moments about the right hand side as follows

-200 (M) +(90 (V)×4) - (4×20)×2 = 0   √

Step 3:

Sum the vertical forces to get V, upwards forces positive

V - (4×20) - 10 = 0

V = 90 kN

4.4   Determination of Internal Responses of Beams to Applied Loads

The internal response (moments and shears) of members to external loads can be calculated once the reactions are known using the three equations of equilibrium.

An alternative way to construct a shear force diagram and to sketch a bending moment diagram showing the maximum moment is as follows.

Shear Force Diagram

First of all draw the length of the beam to scale and chose a suitable vertical scale (eg 1 cm = 10 kN) for the magnitude of the shear force.

Then starting from the left hand reaction follow the load arrows.

Step 1 The reaction force is 28 kN so draw a line 28 units up
Step 2 As there is no load between the left hand support and the centreline draw a horizontal line to the centreline
Step 3 At the centreline there is a 20 kN force down so go down 20 units to 8 kN
Step 4 The uniformly distributed load (UDL) is 36 kN/m so in two metres the line will go down 2×36 = 72 kN to 64 kN.
Step 5 Since the right hand reaction is 64 kN proceed vertically up 64 units (IF YOU DO NOT REACH THE ZERO AXIS YOU HAVE DONE SOMETHING WRONG)

Bending Moment Diagram

The value of the bending moment at a particular point is equal to the area of the shear force diagram to the left of that point. We are usually only concerned with the maximum moment and this occurs where the shear force diagram crosses the x axis.

This point is "x" metres to the right of the centreline. To determine x we start with the value of 8kN at the centreline and if the UDL is dropping at 36kN/m it will take 0.22 metres (8/36) before the shear force is zero.

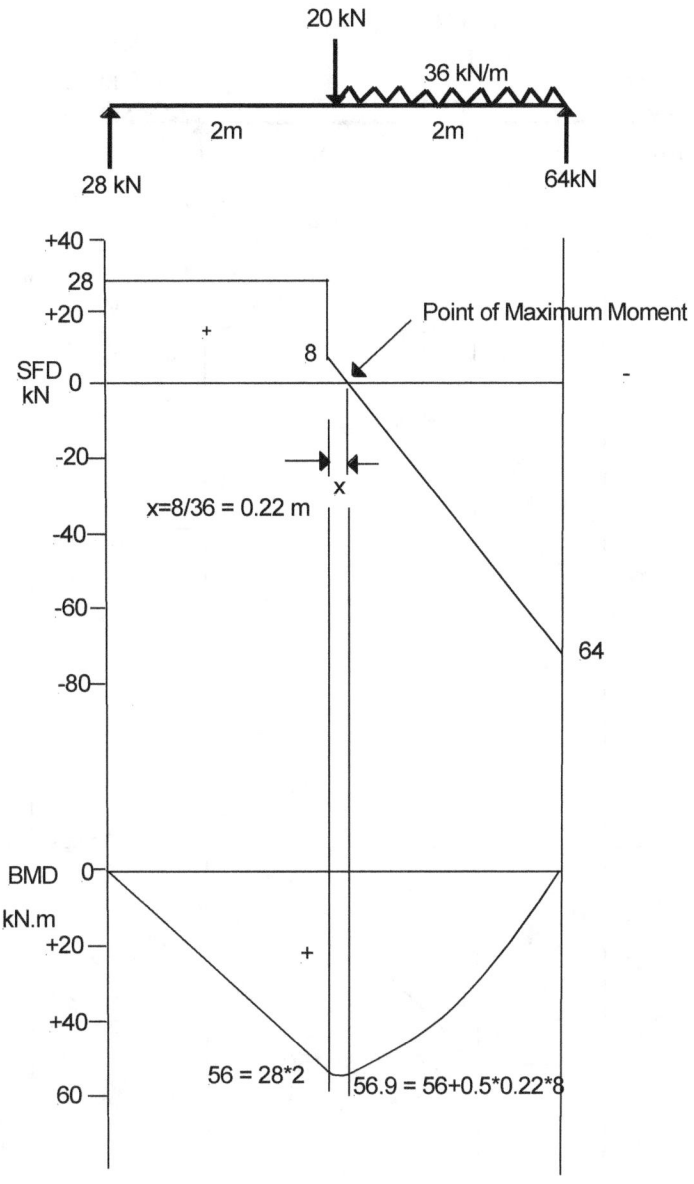

To determine the moment at the centreline we take the area of the shear force diagram to that point (= 28kN × 2 m = 56 kN.m). At the point "x" to the right of this (point of maximum moment the moment will be 56 kN.m plus the area of the triangular portion of the SFD (0.5×0.22×8 = 0.9 kN.m) giving a maximum moment of 56.9 kN.m

The shape of bending moment diagrams are straight lines when there are point loads and concave (downward curving) when there are UDL's so we can roughly sketch the bending moment diagram as follows.

Note: If you have a beam with a cantilever draw the cantilever on the right hand side in your loading sketch.

Now consider the following beam. The reactions have been calculated using the equations of equilibrium. To draw the SFD just follow the arrows.

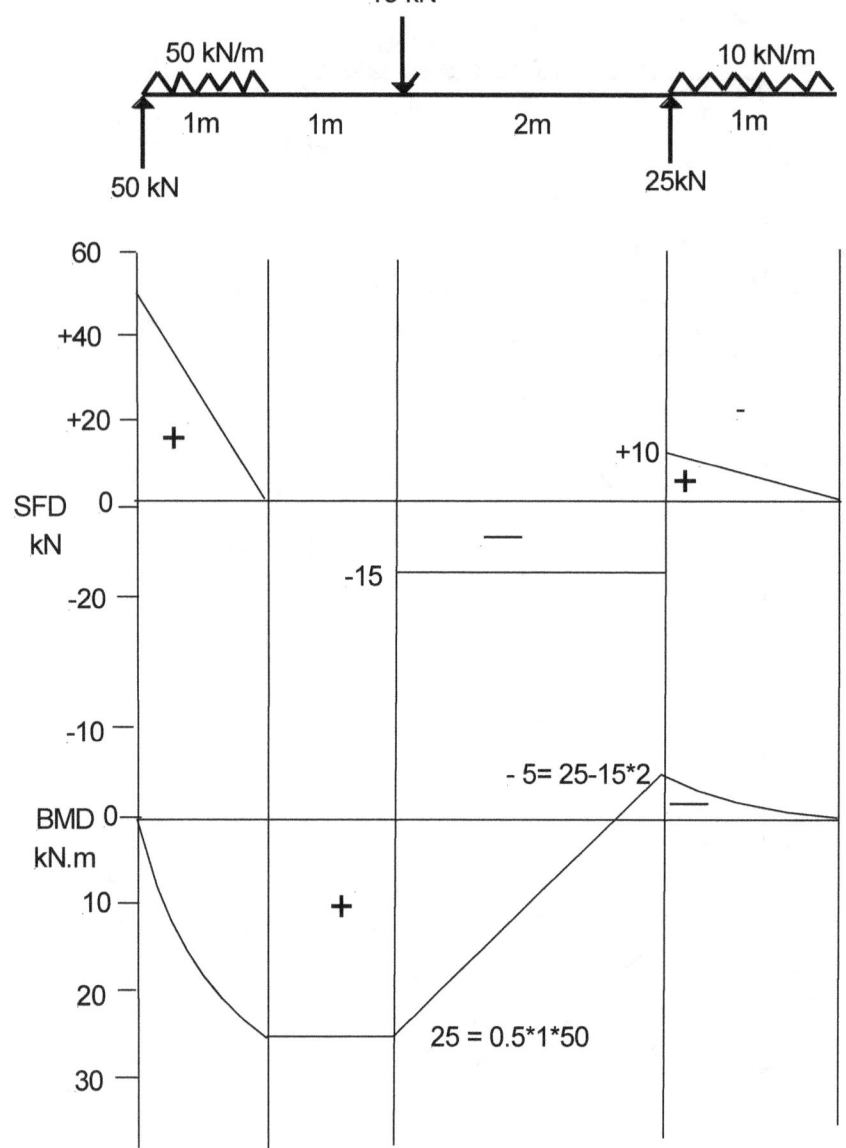

Note that in this case we have two cases where the shear force is zero, one between one and two metres from the left hand support (position of maximum positive moment, i.e. tension on the bottom of the beam) and one at the right hand support (position of maximum negative moment, i.e. tension on the top of the beam) To calculate the maximum positive and negative moments work from left to right.

Maximum positive moment = 0.5×50×1 = 25 kN.m
Maximum Negative Moment = 25 – 15×2 = -5 kN.m

Note that the point where the bending moment changes from positive to negative (point of zero moment) is termed the "point of contraflexure"

4.5 Axial Forces and Stresses in Pinned Frames

For trusses or other frames the analysis of member forces is based on the assumption of pinned joints i.e. each member has no rotational stiffness at the joint. Reactions for such members are determined in the same way as shown for beams above. Analysis then boils down to determining the compressive and tensile forces in the members. This can be done on a computer but if we are only after maximum values then the following simple analyses can be used.

For normal roof trusses the maximum forces in the top and bottom chords usually occurs at the supports and the equilibrium equations can be used to find them as follows.

Consider the following truss

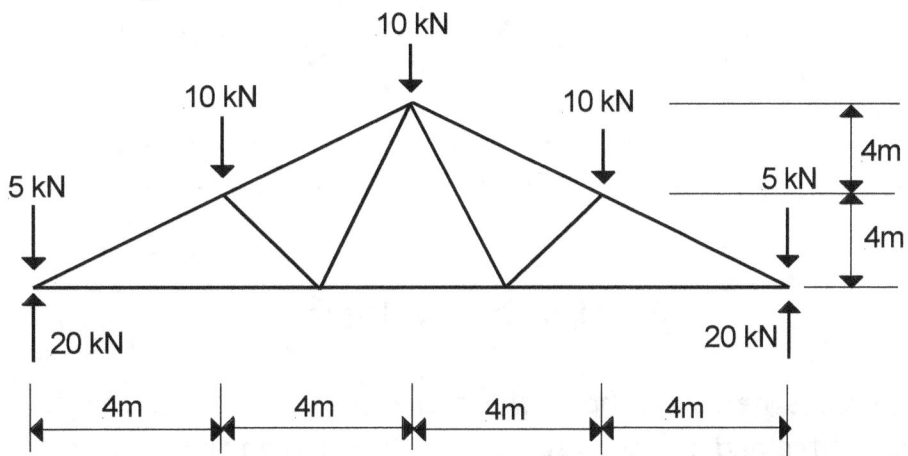

The reactions for this truss can be calculated as 20 kN at each end. The horizontal reaction is zero. The force in the centre of the bottom chord can be obtained by isolating the left hand half of the truss (free body diagram) as follows

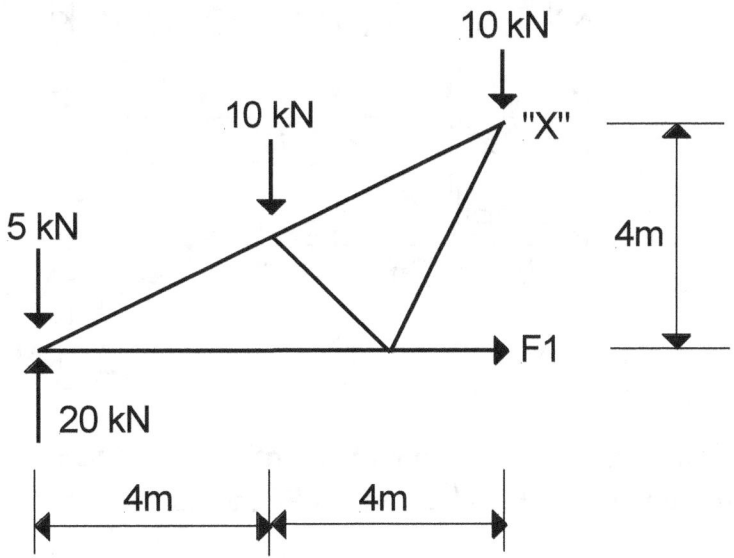

If we take moments about "X" we get

$(-5 \times 8) + (20 \times 8) - (10 \times 4) - (F1 \times 4) = 0$

F1 = 20 kN

Since it is positive it means we have chosen the correct direction of the force, i.e. away from the nearest joint, hence tension

The forces in the web members are harder to get but are generally significantly less that the forces in the top and bottom chords. The full computer generated solution for the above truss is as follows

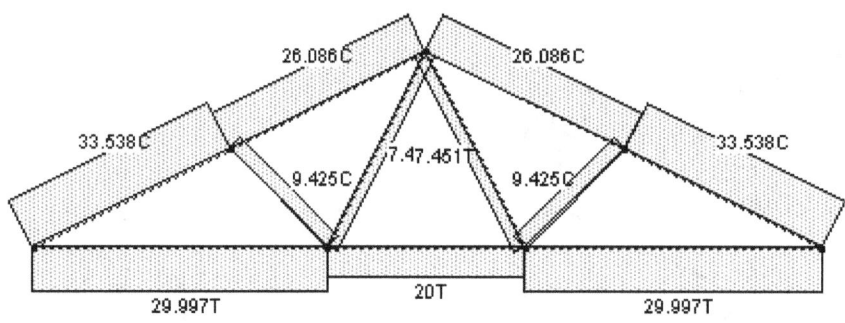

Solution using "Multiframe"

Rectangular trusses generally have their greatest forces in the top and bottom chords at the centre and are analysed as for the sloping truss above.

Consider the following truss

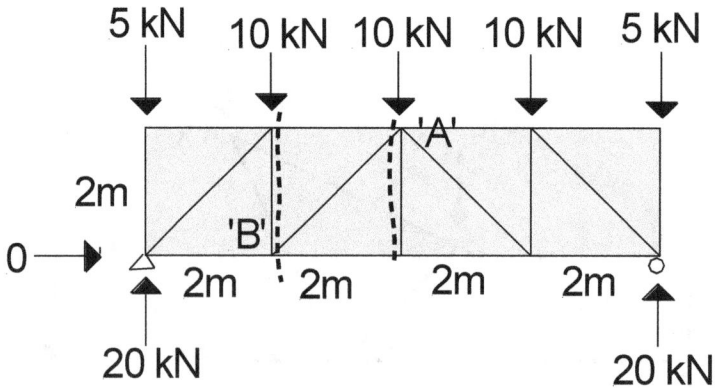

To get the maximum force in the bottom chord at the centre cut the truss just to the left of the central vertical member and take moments about point "A"

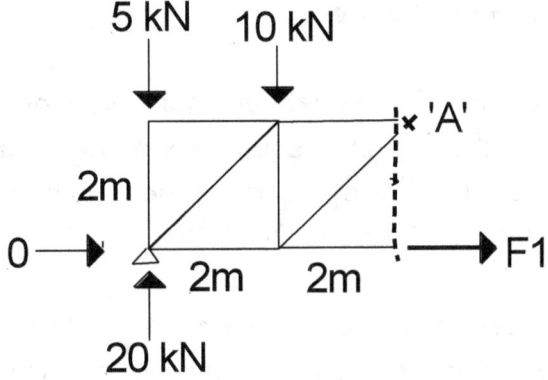

Summing moments about "A" we get

$(20 \times 4) - (5 \times 4) - (10 \times 2) - (F1 \times 2) = 0$
$F1 = 20$ kN (Tension)

To get the maximum force in the top chord at the centre take moments about "B'. Note that the force in the top chord is the same all along to "A"

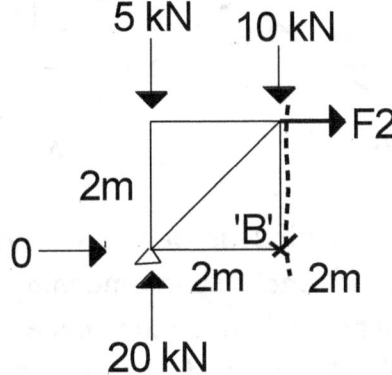

Taking moments about "B" (note that the 10 kN force causes no moment about B, nor does the force in the diagonal emanating from point B) we get

$(20 \times 2) - (5 \times 2) + (F2 \times 2) = 0$

$F2 = -15$ kN (minus means it should be into the joint, therefore compression)

The "Multiframe" computer generated solution is given below

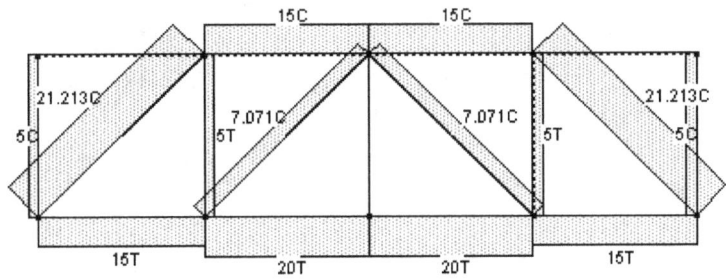

For a graphical method of determining forces in trusses see Appendix D.

4.6   Reactions of Continuous Beams

Continuous beams are beams with three or more supports.   As such they have greater than 3 reactions and the three equations of equilibrium are insufficient to calculate the reactions exactly.   Exact hand methods are available as are computer solutions using packages such as "Multiframe"

Continuous beams have both positive (tension on bottom of beam) and negative (tension on top of beam) moments, as illustrated in the diagram below.   The point where tension changes from being on the bottom of the beam to being on the top of the beam (where the bending moment is zero) is termed the "point of contraflexure".

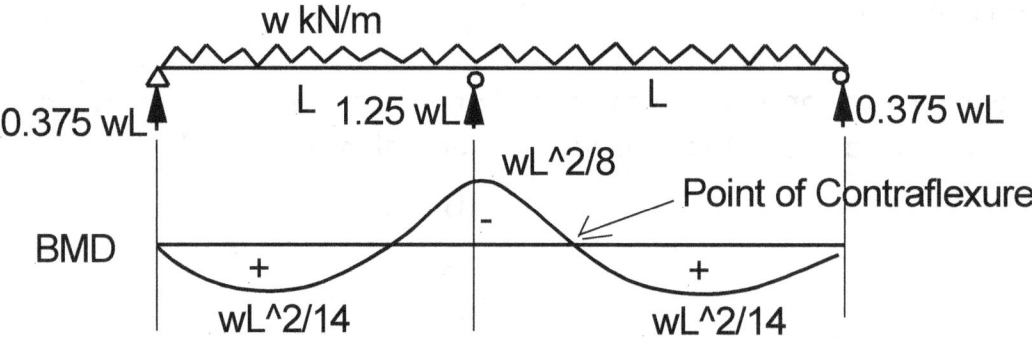

The other point to note about continuous beams is that the reactions at the ends are less than half the load on each span, meaning that more of the load is transferred to the interior supports (the shear force adjacent to interior supports is consequently also higher.   In the case of the two span continuous beam shown it is 0.625w×L rather than 0.5w×L which would be the case for a simply supported span.

4.7   Reactions in Portal Frames

Generally speaking portal frames are considered pinned at their bases and therefore have four reactions – a horizontal and a vertical at each column base.   By assuming that the two horizontal reactions are equal we can reduce the unknowns to 3 and apply the three equations of equilibrium as for beams.

For example

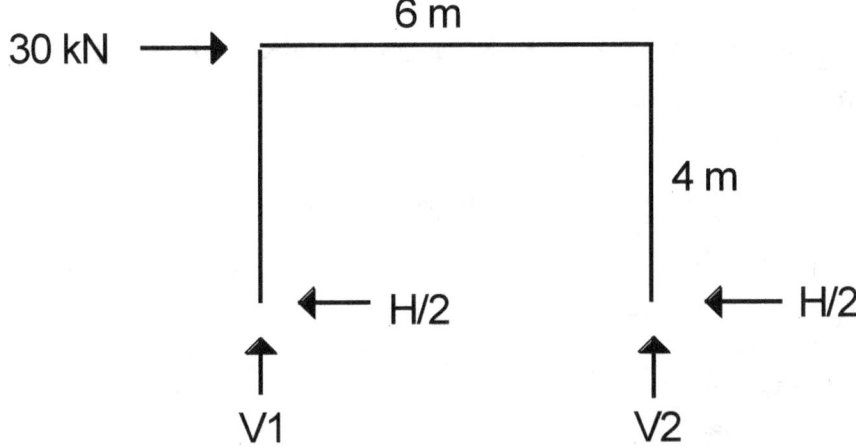

Summing horizontal forces gives

+30 –H/2 –H/2 = 0

H = 30 KN

Taking moments about the left hand side gives

(30×4)-V2×6 = 0

V2 = + 20 kN

Taking moments about the right hand side

(V1 ×6)+ (30 ×4) = 0

V1 = -20 kN (The minus sign means we should have drawn the reaction acting down)

Bending Moment diagrams for frames are best done by computer. The following is a BMD for a simple frame with a horizontal load. Note that moments are drawn on the face that will be in tension under the load. So in this case the inside face of the left haunch will be in tension and the outside face of the right, as we might expect when a horizontal load is applied to the left hand haunch.

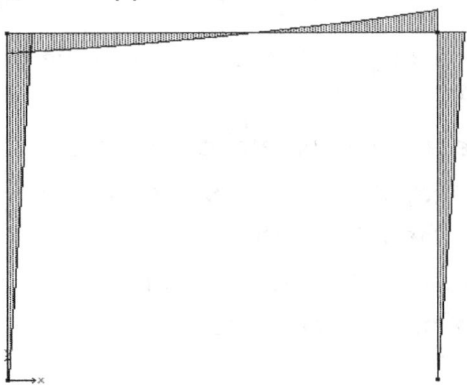

Moments are generally largest at the haunches which is why industrial portal frames have thickened sections at the haunch.

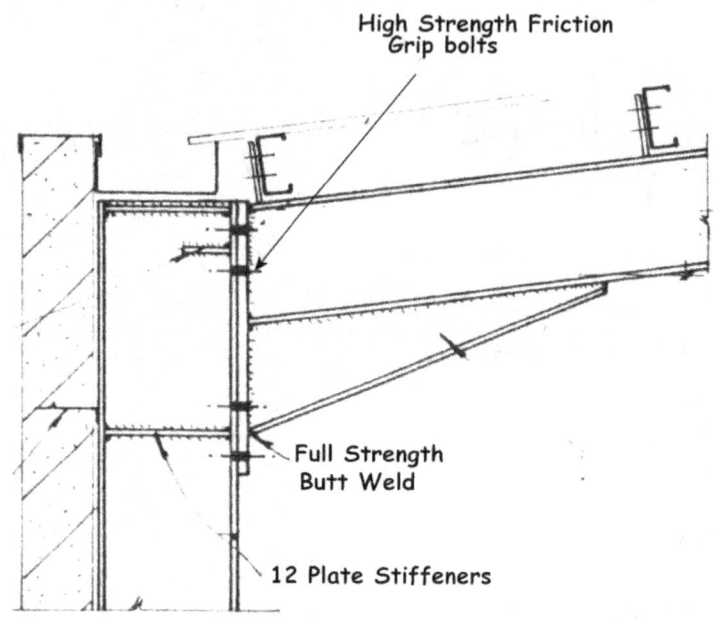

Figure 4.2   Knee Haunch in Portal Frame

4.8   Deflection of Beams and Frames

4.8.1   Beams and Frames

The deflection of beams and frames is more often than not the limiting factor in the design of permanent structures.   For temporary structures deflection of elements may not be critical in the case of such things as strutted excavations but is certainly critical in cases such as formwork.

The deflection of a beam or frame is dependant on a property called "bending stiffness".   The bending stiffness of a beam or frame varies depending on support conditions but is always a multiple of $E \times I$ where

I = Moment of inertia of member (mm^4) = measure of how stiff the beams shape is (See Appendix C3)
   $= 1/12 \times b \times d^3$ for rectangular members of width b and depth d
   = Values given in steel manufacturer's brochure (Generally I_{xx})
      for steel members (See Appendix C)
   $\approx 0.045 \times b \times d^3$ for reinforced concrete beams
   $\approx 0.033 \times b \times d^3$ for reinforced concrete slabs

In general the more part of a section's area is away from the centroid (centre of mass or centre of gravity) the greater will be its moment of inertia. That is why steel I beams are much stiffer than would be rectangular beams of equivalent area.

E = Modulus of Elasticity of Material (MPa) – measure of how elastic a beams material is, i.e. how much it stretches for a given tension load
    = 200,000 MPa for steel
    = 5055 × $\sqrt{}$ F'c Reinforced Concrete
    = 26,875 MPa for F'c = 25 MPa
    ≈ 70,000 MPa for aluminium
    = 12,000 for F14 Timber (Some Hardwoods)
    = 10,500 for F11 Timber (Unidentified hardwoods)
    = 9,100 for F8 Pine
    = 7,900 for F7 Oregon
    = 6,900 for F5 Pine

The deflection of beams depends on the total load (W = w × L or P), the span of the member (L) and its stiffness (EI). In general the formula for beams is

$$\text{Deflection} = \text{Constant} \times \text{Load} \times L^3 / (EI)$$

From this one can see for example that if the load and stiffness remains constant then doubling the span will effectively increase the deflection by a factor of 8 (2^3)

Long term deflection of beams is important in permanent designs and for temporary structures which remain in place for a long time. In general long term deflections for steel and aluminium beams will be equal to the short term deflection. For seasoned timber beams long term deflection will be about twice the short term deflection and for unseasoned timber three times. For reinforced concrete elements long term deflections will be around two and a half times the short term deflection.

Section 4.9 gives some standard solutions for deflection of beams.

4.8.2   Trusses

The deflection of trusses depends on the "axial stiffness" of members. Axial stiffness is a function of E×A , where E = Modulus of elasticity of the material (MPa) and A is the cross-sectional area of the member in mm^2.

4.9 Standard solutions for Reactions, Max Shear Force, Max bending Moment and Deflection

Beam Type	Reactions (kN)	Maximum Shear Force (kN)	Maximum bending Moment (kN.m)	Maximum Deflection (mm) SEE NOTE BELOW
w (kN/m) L (m) S.S. BEAM WITH UDL	$\dfrac{w \times L}{2}$	$\dfrac{w \times L}{2}$	$\dfrac{w \times L^2}{8}$	$\dfrac{5 \times w \times L^4}{384 \times E \times I}$
P (kN) L (m) S.S. BEAM WITH POINT LOAD IN CENTRE	$\dfrac{P}{2}$	$\dfrac{P}{2}$	$\dfrac{P \times L}{4}$	$\dfrac{P \times L^3}{48 \times E \times I}$
w (kN/m) w (kN/m) L (m) L (m) TWO SPAN CONTINUOUS BEAM WITH UDL	$\dfrac{3 \times w \times L}{8}$ Ends $\dfrac{10 \times w \times L}{8}$ Centre	$\dfrac{5 \times w \times L}{8}$	$-\dfrac{w \times L^2}{8}$ Over Support $+\dfrac{w \times L^2}{14.22}$ Midspan	$\dfrac{w \times L^4}{185 \times E \times I}$
Right Reaction = -3P/32 P = = L (m) L (m) TWO SPAN CONTINUOUS BEAM POINT LOAD	$\dfrac{13 \times P}{32}$ L.H. End $\dfrac{11 \times P}{16}$ Centre	$\dfrac{19 \times P}{32}$	$-\dfrac{3PL}{32}$ Over Support $+\dfrac{13PL}{64}$ Midspan	$\dfrac{3 \times P \times L^3}{200 \times E \times I}$
w (kN/m) L (m) CANTILEVER BEAM WITH UDL	$w \times L$	$w \times L$	$\dfrac{w \times L^2}{2}$	$\dfrac{w \times L^4}{8 \times E \times I}$
P (kN) L (m) CANTILEVER BEAM WITH POINT LOAD AT END	P	P	$P \times L$	$\dfrac{P \times L^3}{3 \times E \times I}$

Note for Deflection Calculations:

For UDL's w must be in N/mm (=kN/m), L in mm, I in mm^4 and E in MPa
For Point Loads P must be in N, L in mm, I in mm^4 and E in MPa

For beams with combinations of loads the results can be calculated separately and added together to get the final answer.

Consider the following 300 by 100 F7 beam (E=7,900 MPa)

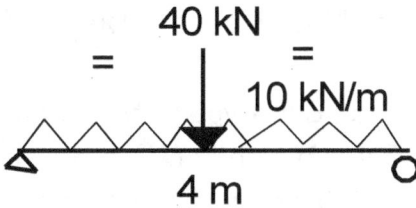

40 kN

10 kN/m

4 m

$I = BD^3/12 = 100 \times 300^3/12 = 225 \times 10^6 \text{ mm}^4$

This can be broken down into the following two beams

Beam 1

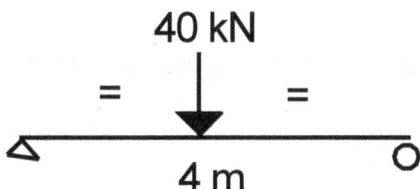

10 kN/m

4 m

Reactions $= \frac{w \times L}{2} = 10 \times 4/2 = 20$ kN

Max Shear Force $= \frac{w \times L}{2} = 10 \times 4/2 = 20$ kN

Max Bending Moment $= \frac{w \times L^2}{8} = 10 \times 4^2/8 = 20$ kN.m

Maximum Deflection $= \frac{5 \times w \times L^4}{384 \times E \times I} = (5 \times 10 \times 4000^4)/(384*7900 \times 225 \times 10^6)$

$= 18.8$mm

Beam 2

40 kN

4 m

Reactions $= \frac{P}{2} = 40/2 = 20$ kN

Maximum Shear Force $= \frac{P}{2} = 40/2 = 20$ kN

Maximum Bending Moment $= \frac{P \times L}{4} = 40*4/4 = 40$ kN.m

Maximum Deflection $= \frac{P \times L^3}{48 \times E \times I} = (40,000 \times 4000^3)/(48 \times 7,900 \times 225 \times 10^6)$

$= 30$ mm

Final Result:

Reactions $= 20 + 20 = 40$ kN

Maximum Shear Force $20 + 20 = 40$ kN

Maximum Bending Moment $= 20 + 40 = 60$ kN.m

Maximum Deflection $= 18.8 + 30.0 = 48.8$ mm

4.10 Analysis of 2D Structures using "Multiframe"

The "Multiframe" structural analysis program (www.formsys.com.au) has 6 windows but you can basically get by with the "Frame", "Load" and "Plot windows. Start a new file then make sure that you have Australian units (View/Units) and that you have selected "Load values" and "Plot values" in Display/Symbols

Frame Window– here is where you draw your structure and define members. First of all set a grid scale in View Grid, typically 0.5 m or 1 m.   Then choose Frame/Add Member and draw your frame. Select each member and assign a section type in Frame/Section Type . Select each point where the frame is restrained (bases of portal frames or reaction points of beams) and allocate a restraint in Frame/Joint Restraints . Joint restraints may be either pinned , fixed or roller. Simply supported beams have one pinned and one roller restraint, continuous beams one pinned and the rest rollers and portal frames have all column bases pinned.

Load Window– here is where you apply loads to the structure. For loads applied at joints choose Load/Joint Load, for loads applied along member choose either Load/Global Distributed Load or Load/Global Point Load

Once these two windows are complete the structure is analysed (Case/Analyse) and then results are viewed in the "**Plot**" window. Use Display/Actions for Moment (M_z), Shear Force(V_y) and Axial Force (For trusses P_x). Use Display/Deflection for deflections. To see moments shears etc along an element select the element and move the cursor along.

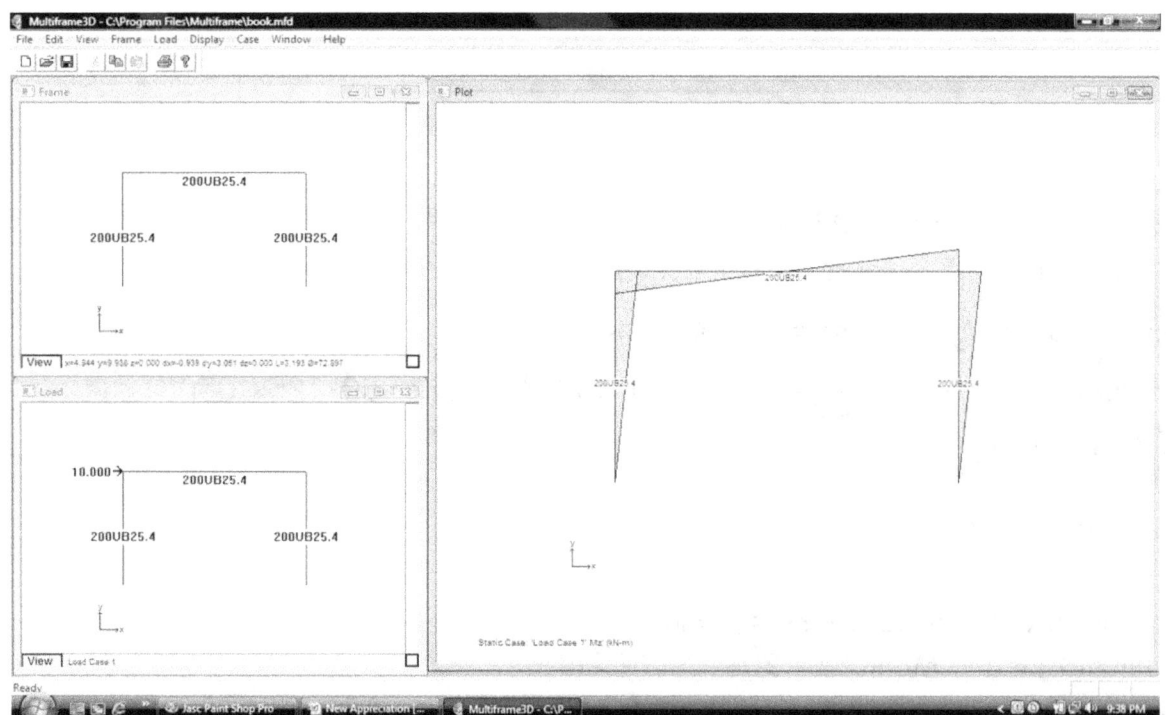

Figure 4.3 Windows from Multiframe Program

4.11 Arches and Cable Structures

Arches are similar to cables in that the shape a cable adopts under load is equal to the "thrust" line in an arch.

Figure 4.4 below shows the different shapes cables adopt depending on their load. Under self weight only a cable will fall into a catenary shape but with uniformly distributed load (such as occurs in a bridge) the shape is parabolic. There is however not much difference between these two shapes if the sag is small.

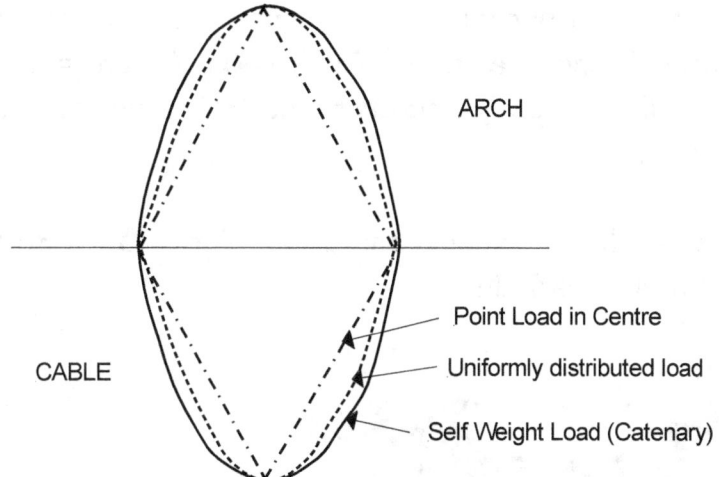

Fig 4.4   Cable Shapes and Arch Thrust Lines

The thrust line in arches is important because if the thrust line falls within the middle third of the arch section then the arch will be in compression only. Stone arches cannot take any tension because of the joints.

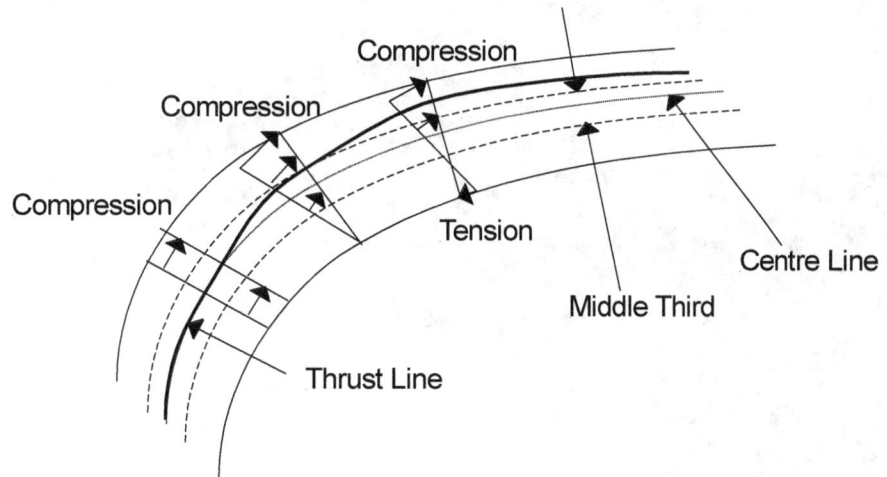

4.12.1   Finding Compressive Force in Arches

To find the compressive force in an arch first of all calculate the vertical reaction based on the loads. For example if we have a pathway sitting on top of a 6m arch with a load of 1.5 kN/m then the reactions at each end are 4.5 kN (1.5×6/2).

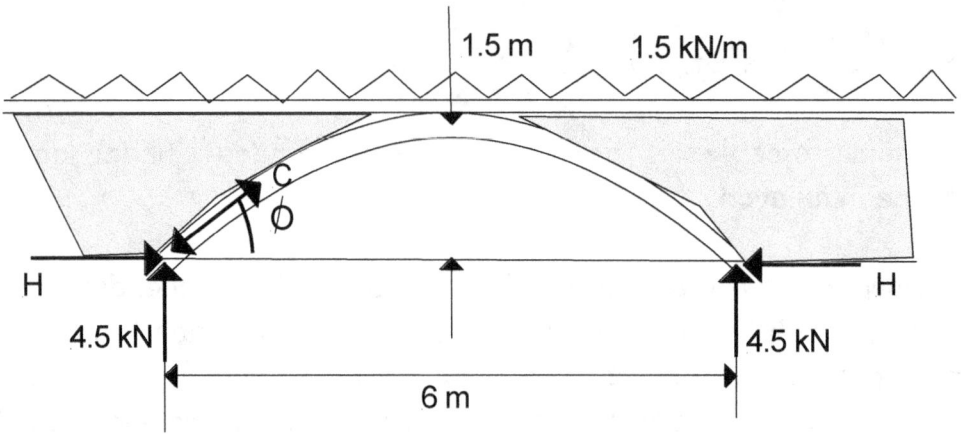

Assuming that the arch is a parabola with a rise of 1.5 metres and a span of 6 m then the angle Ø at each end is equal to $\text{Tan}^{-1}(4\times\text{Rise}/\text{Span}) = \text{Tan}^{-1}(4\times1.5/6) = 45°$. Let C be the compressive force in the arch and H the horizontal reaction needed to keep the arch in shape.

Then Vertical Force (4.5 kN) = C×CosØ and C = 4.5/Cos 45 = 6.36 kN
H = C×SinØ = 6.36×Sin 45 = 4.5 kN

Figure 4.5   The Carrick-A-Rede Rope Bridge in Northern Ireland

EXERCISES:

1. Calculate the reactions of the following beams (Answers given) and then draw shear force and bending moment diagrams.

a)

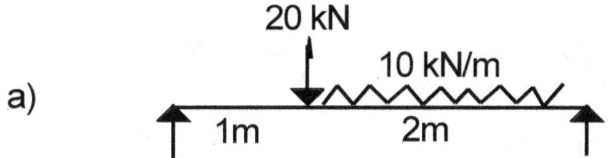

b)

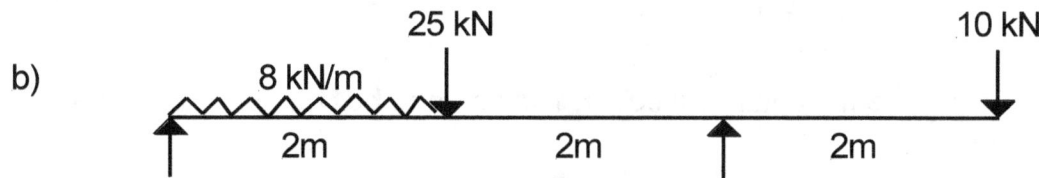

c)

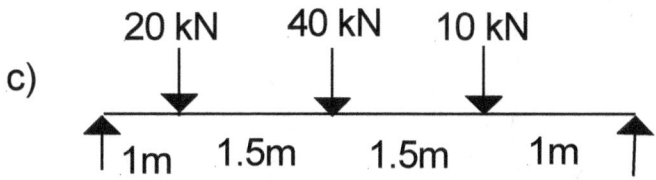

d)

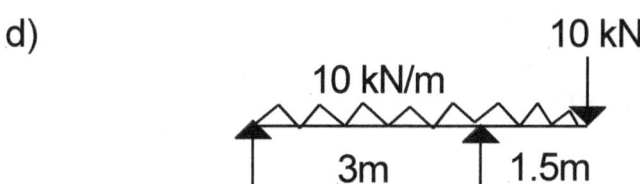

e)

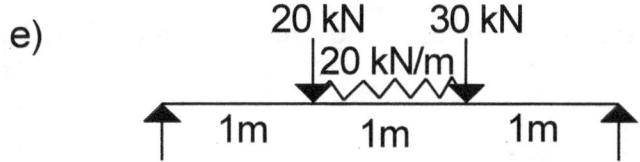

2. Determine the maximum forces in the top and bottom chords of the following truss

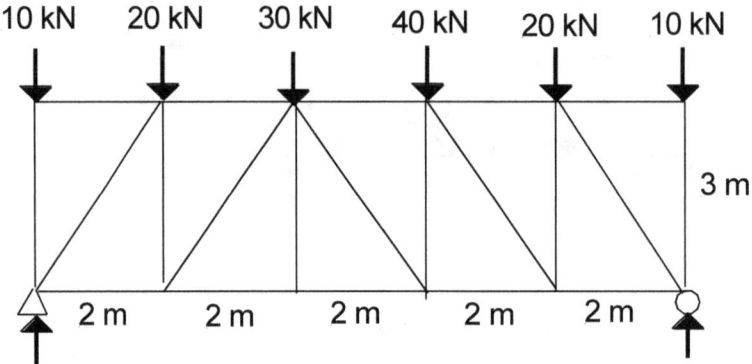

3. Determine the maximum forces in the top and bottom chords of the following truss

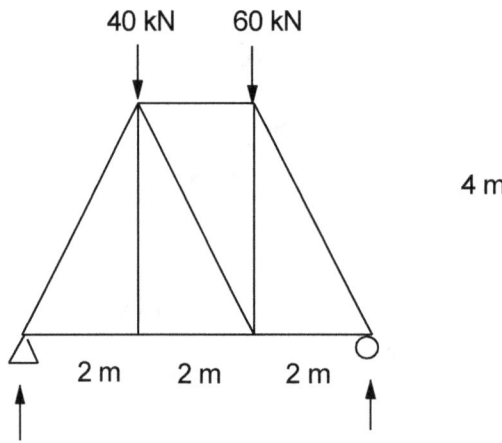

4. Calculate the maximum deflection of the following steel beam (310UB40 – $I_{xx} = 85.2 \times 10^6 mm^4$

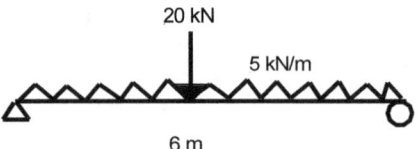

5. DETERMINATION OF STRUCTURAL SAFETY

5.1 Factors of Safety

In determining the safety of a structure or structural element the question that must be asked is

> ### HOW IS IT POSSIBLE FOR THIS STRUCTURE TO FAIL?

Once this is determined we can then proceed to work out the loads and check whether the structure or structural element has a sufficient factor of safety against this type of failure.

The most common ways a structure or structural element can fail is by

- Breaking due to bending or shear in flexural members, crushing due to compressive forces in compression members or snapping due to tensile forces in tension members (strength failure). This includes failure due to local member instability such as buckling of columns or lateral buckling of top flanges in beams
- Excessive deflection (serviceability failure)
- Overall instability failure, i.e. by uplift, sliding or turning over.

5.2 Strength Failure

In the old days the strength safety of a structural element was determine using "allowable stresses", i.e. the calculated stresses in the element were required to be less than the failure stresses of the material divided by a "Factor of Safety". The loads used to calculate the stresses were the actual loads imposed on the structure. This approach is still used nowadays for checking the bearing pressures underneath footings and comparing these with the "Allowable Bearing Pressure" (See Section 3.6)

Nowadays we use a "load factor" approach for the design of structural elements such as beams, slabs or columns. **In the Load Factor approach the actual loads (eg gravity loads) are scaled up by "Load Factors" and the resultant load is called the "Factored Load".** Structural responses to the factored loads (termed Member Actions) are then calculated as detailed in Chapter 4. For example N_c^* refers to the compressive force in a member due to factored loads.

For strength checks on structural members Australian code AS 1170.1 requires you to consider combinations of factored loads and to design for the worst case. The main combinations are as follows

1.2 × Dead Load + 1.5 × Live Load
1.35 × Dead Load
1.2 * Dead Load + Wind Load + 0.4 * Live Load

Earth loads are treated as Live Loads . **Note also that wind pressures (based on V_u) are already "factored" in the Wind Code.**

We define the failure resistance of an element as the load at which the element breaks (this may be determined by test or by theory). The failure resistance is scaled down by a "Capacity Reduction Factor" (ϕ) which is related to the variability of the particular material strength to give "<u>Member Capacities</u>". For example ϕN_c denotes the compressive capacity of a column.

The design process is expressed mathematically as follows

$$\phi W \text{ must be greater than } W^*$$

where

ϕW= Element Capacity

W^* = Factored Action within the element due to factored loads

For example in the above case ϕN_c must be greater than N_c^*.

The value of ϕ to be used will depend on the material being considered and on the type of failure being checked eg bending or shear. The following values will be assumed in this book

For steel members ϕ = 0.9

For concrete members ϕ = 0.8

For timber members which can be classified as "primary members", i.e ones for which failure would be catastrophic (eg columns) ϕ = 0.65

For other timber members ϕ = 0.8

In essence the Load Factor procedure is equivalent to defining a "Factor of Safety" as equal to Load Factor /ϕ. Thus for steel members an equivalent Factor of Safety would be 1.5/0.9 = 1.67 for live loads and 1.2/0.9 = 1.33 for dead loads.

<u>Note however that for foundation design there is no current load factor code and the allowable stress approach is still used</u> . When checking foundation bearing pressures we still calculate the pressure due to unfactored loads and compare this with the Allowable Bearing Pressure. This is particularly annoying since for example we must use factored loads to check columns above the footing but unfactored loads to check foundation pressures.

5.3 Serviceability Failure

Serviceability failure refers to the condition where a structure is considered to be "unserviceable" because the forces acting on it produce visually or psychologically unacceptable consequences. Examples of serviceability failure are excessive deflection of beams or excessive lateral movement in tall buildings or excessive cracking in slabs.

Serviceability loads are "unfactored" or "actual" loads and the limits determining serviceability failure are subjective. We will assume a deflection limit of Span/300 for beams and slabs subjected to combinations of dead and live loads. For example if a beam is 6 metres long the allowable deflection would be 6000/300 = 20 mm.

For wind loads the serviceability wind speed V_s is based on a return period of 20 years whilst V_u (as used in strength calculations) is based on a 500 year return period. The ratio of the wind pressures calculated using V_u and V_s is around 1.5, equivalent to the wind load factor.

5.3.1 Deflection Calculations

Most of the deflection calculations done in this book relate to flexural members, i.e. beams or slabs. The resulting deflections should be compared with the allowable deflection of Span/300.

5.3.1.1 Point Loads

In calculating deflections due to point loads the full load should be used. If the point load is to be there for a long time the appropriate creep factor should be applied if the member is timber or concrete (See below)

5.3.1.2 Uniformly Distributed Loads (UDL)

For UDL's (kN/m or kN/m^2) **short term deflections** will be based on the dead load plus 70% of the live load. 70% of the live load is used rather than the total live load as this represents the "most likely" load rather than the maximum load.

In this book the calculation of **long term deflections** for UDL's will be based on dead loads plus a proportion of the underlined uniformly distributed live load which may be assumed to be long acting (usually taken as 33%).

(Note that if the UDL is a wind load the full load should be used).

5.3.1.2   Long Term Creep Factors

In timber and reinforced concrete beams and slabs long term deflections are much greater than initial deflections due to creep, and the calculated deflections **due to long term loads** should be multiplied by a long term creep factor to obtain the maximum deflections.

For timber

long term creep factor (j_2) = 2 for seasoned timber

= 3 for unseasoned timber

For reinforced concrete beams and slabs

long term creep factor = $[3 - 1.2 \times A_{sc}/A_{st}]$ where Asc is the cross sectional area of steel in the compressive zone of the beam (top for simply supported members) and Ast is the cross sectional area of steel in the tension zone.

long term creep factor = Min 1.8 for $A_{sc} \geq A_{st}$

long term creep factor = 2 if $A_{sc} = 0.83 \times A_{st}$

long term creep factor = 3 if $A_{sc} = 0$

5.4   Overall Instability Failure

In permanent structures problems of overall stability occur in many cases, the more common being
* Overturning of Portal Frames due to wind
* Overturning of Retaining Walls due to soil pressure
* Uplift of roofs due to wind
* Overturning of cranes or lifting frames due to loads being lifted.
* Failure of soil slopes (slip circle failure)

Overall instability of structures needs to be checked based on either of the three basic equations of equilibrium, depending upon what is causing the instability.

1. Factored <u>horizontal</u> forces causing instability must be less than the factored resisting forces OR
2. Factored <u>vertical</u> forces causing instability must be less than the factored resisting forces OR
3. Factored moments causing instability must be less than the factored resisting moments.

For the definition of factored loads see Section 4.2.2. Note
1. When dead loads are the primary loads causing instability they are factored by 1.35.
2. When there is a combination of dead and live loads the dead loads are factored by 1.2 and the live loads by 1.5. However if the dead loads factored by 1.35 produce a higher destabilising effect then this is the situation to consider.
3. When wind loads are the primary loads causing instability the unfactored "strength" wind pressures should be used. When there is a combination of dead and wind loads the dead loads are factored by 1.2. However if the dead loads factored by 1.35 produce a higher destabilising effect then this is the situation to consider
4. Dead loads resisting instability are factored by 0.9.
5. Lateral soil loads are factored by 1.5.

Example 1:

Consider the case of a man (1kN) standing on the end of an overhanging plank (2 metres overhang) with a concrete block counterbalance (3.75 kN) 1 metre from the edge.

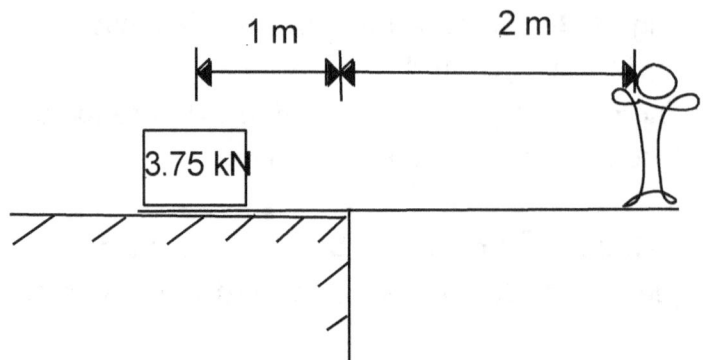

The calculations for checking overall stability about the edge would look like this.
Factored Overturning Moment = 1.5 (Live Load Factor) × 1kN × 2 metres = 3 kN.m
Factored Resisting Moment = 0.9(Load Factor) × 3.75kN × 1 metre = 3.38 kN.m

Plank is stable since factored overturning moment is less than the factored resisting moment.

Example 2:

Consider the stability of the following sign. Assume all parts of the sign except the concrete bases are weightless.

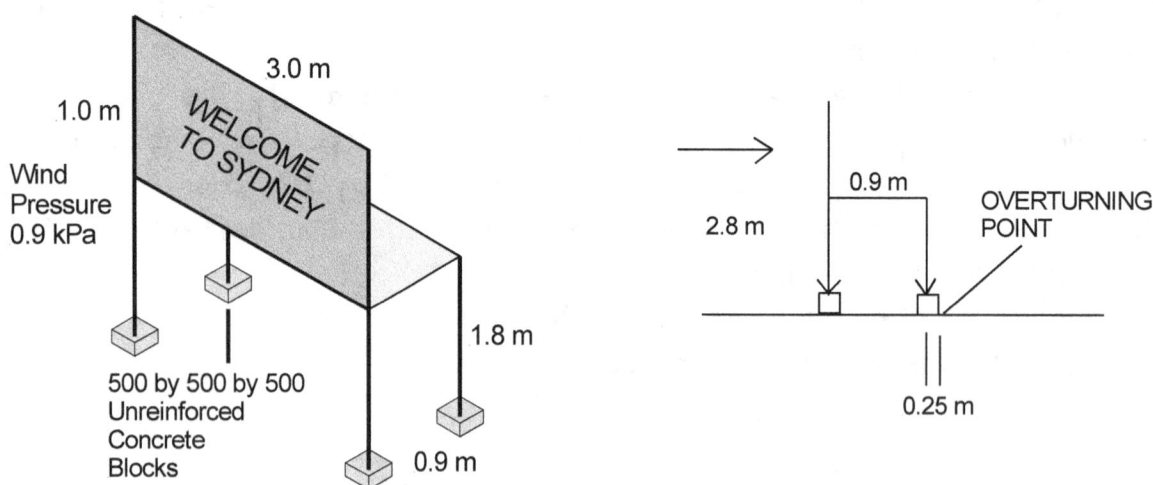

The overturning moment ($M_{O/T}$) is equal to the wind force times the distance between the centre of the sign and the overturning point (0.5+1.8+0.5 = 2.8 m)

$M_{O/T}$ = (3×1×0.9)×2.8 = 7.56 kN.m

The resisting moment (M_{RES}) is equal to the gravitational force of the outer two blocks (reduced by the load factor of 0.9) times the distance between the blocks and the rotation point (back edge of inner blocks) plus the gravitational force of the inner two blocks (reduced by the load factor of 0.9) times the distance between these blocks and the rotation point.

M_{RES} = [(2×0.5×0.5×0.5×24×0.9)×(0.9+0.25) + (2×0.5×0.5×0.5×24×0.9)×0.25]
        = 7.56 kN.m

Since the overturning moment ($M_{O/T}$) is less than or equal to the resisting moment (M_{RES}) the structure is stable.

Example 3:

Consider the case of the reinforced concrete retaining wall shown below. To check its overall stability we first need to establish a point on the structure about which the structure would rotate if it overturned. This is done by inspection and in this case we could imagine the wall overturning about its toe. Note that other possible points should also be checked

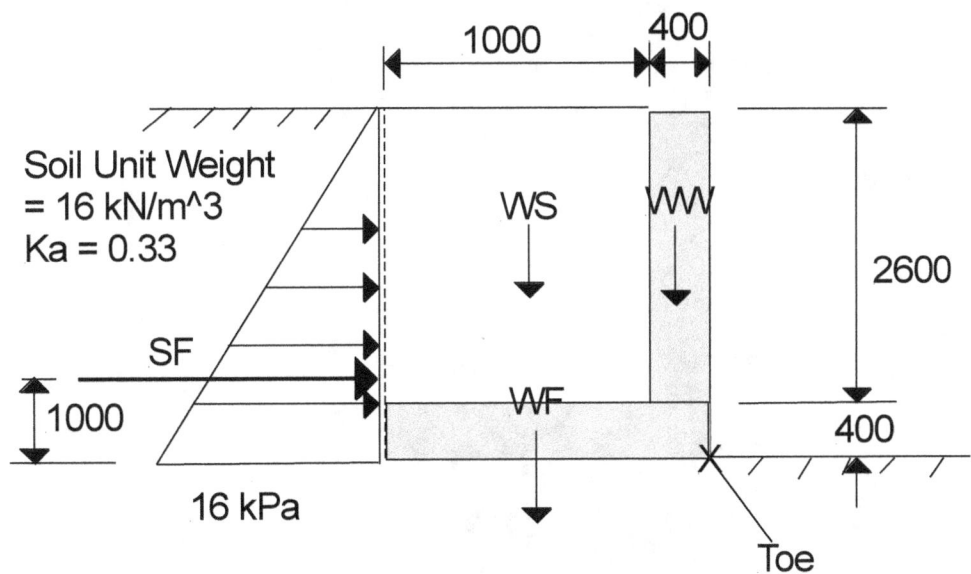

W_W = Weight of wall = 0.4 × 2.6 × 25 = 26.0 kN/m
F_W = Footing Weight = 1.4 × 0.4 × 25 = 14.0 kN/m
S_W = Soil Weight = 2.6 × 1.0 × 16 = 41.6 kN/m
Soil Pressure = Ka×γ×H = 0.33×16×3 = 16 kPa
Soil Force S_F = 0.5×16×3 = 24 kN
Moments Resisting Overturning

The resisting moments are to be multiplied by a load factor of 0.9 and are
Due to W_W = 0.9 × 26.0 × 0.2 = 4.7 kN.m

67

Due to $F_W = 0.9 \times 14.0 \times 0.7 = 8.8$ kN.m
Due to $S_W = 0.9 \times 41.6 \times (0.5 + 0.4) = 33.7$ kN.m
Total Resisting Moment = 4.7 + 8.8 + 33.7 = 47.2 kN.m / m

Moments Causing Overturning

Soil loads have a load factor of 1.5 so in this case the overturning moment of the soil force about the toe = $1.5 \times 24.0 \times 1.0 = 36$ kN.m / m. (Note that the centroid of a triangular distributed pressure is at the third of its height, i.e. one metre from the base)

As the resisting moment is greater than the factored overturning moment the structure is safe.

EXERCISES:

1. Check the overall stability of the following brick retaining wall. Assume soil backfill unit weight of 18 kN/m^3 and K_a = 0.33.

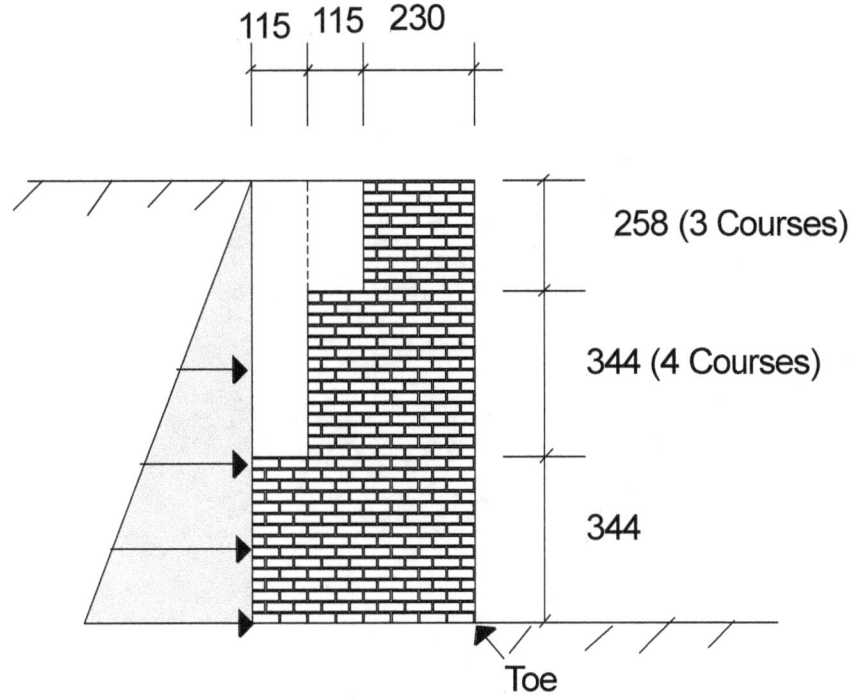

115 115 230

258 (3 Courses)

344 (4 Courses)

344

Toe

2. Check out the overall stability of the following crane. Include impact factor of 1.25 for load of crane (10 tonnes).

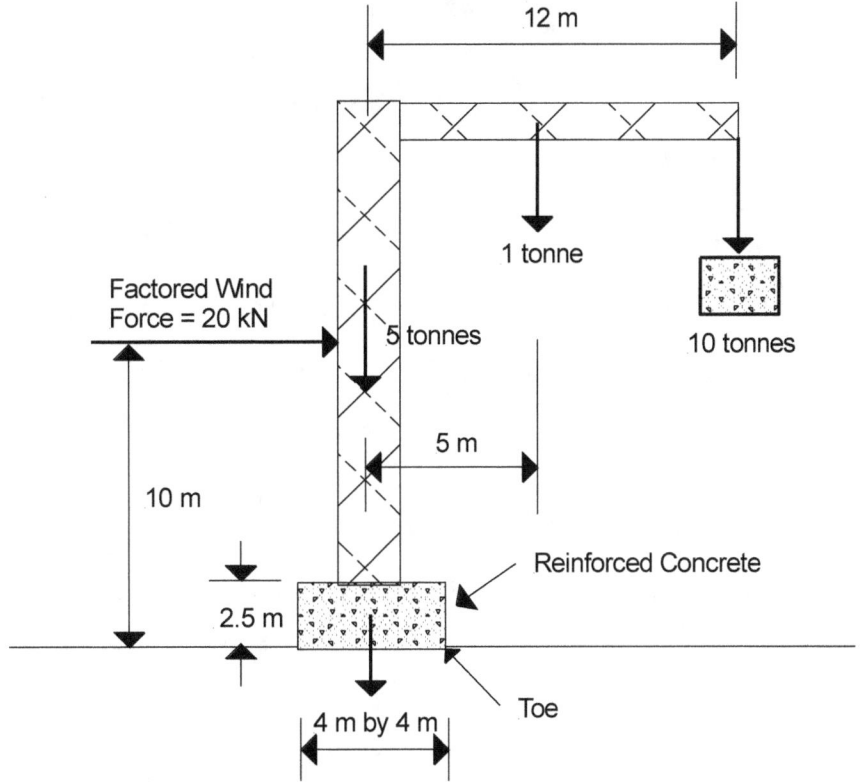

12 m

1 tonne

Factored Wind
Force = 20 kN

5 tonnes

10 tonnes

5 m

10 m

Reinforced Concrete

2.5 m

Toe

4 m by 4 m

6. DESIGN OF SIMPLE TIMBER STRUCTURAL ELEMENTS

6.1 General

This chapter is an attempt to present the basic concepts of structural design as applied to timber elements. It will aim at illuminating the more important variables and will approximate the lesser ones with constant factors. These constant factors will be chosen to give results in practical situations which are not too different from those which would occur if the more accurate methods were adopted. They may however be reasonably conservative in some cases but this is thought not to be a disadvantage in view of the uncertainty regarding calculation by non-engineers of the loads to be resisted.

The principle approach to be adopted in this chapter is what is loosely called the *Limit State* approach. Put simply there are two limit states to be considered in designing structural elements.

6.2 Limit State for Strength

What is required here is that the capacity of the element (in compression, tension, bending and shear) be less than the calculated forces in the element due to the applied loads factored by appropriate safety factors. In the Australian situation the appropriate safety factors are 1.20 for dead loads and 1.50 for live loads however this distinction is not immutable and in some cases it may be appropriate to use a factor of 1.50 for dead loads when the load path is not easily defined.

The capacity of an element for the above strength states - compression/tension, bending and shear - is primarily dependent on the yield stress of the material (stress at which the material ceases to behave elastically) and is reduced by a capacity reduction factor called ϕ to provide a further factor of safety. This factor ϕ takes into account the accuracy of the relevant equation in predicting the response of the material. For timber members which can be classified as "primary members", i.e ones for which failure would be catastrophic (eg columns) $\phi = 0.65$. For other timber members $\phi = 0.8$

The strength limit state criteria are summarised by the following equations.

$$\phi\, N_t \geq N_t^* \quad \text{for Tension}$$

$$\phi\, N_c \geq N_c^* \quad \text{for Compression}$$

$$\phi\, V_v \geq V^* \quad \text{for Shear}$$

$$\phi\, M_b \geq M^* \quad \text{for Bending}$$

where N_t^* etc are the calculated forces/moments in the member resulting from the factored loads and ϕN_t etc are the calculated abilities of the element to resist such actions.

6.3   Limit State for Serviceability

Although serviceability covers many other issues such as durability the principle serviceability concern in most cases is deflection.   The limit state for deflection may be thought of as occurring when the deflection of an element reaches the desirable maximum (allowable) deflection. In terms of design this means that the calculated deflection of the element based on <u>unfactored</u> or actual loads must be less than a specified allowable deflection.

$$\delta_{CALCULATED} \leq \delta_{ALLOW}$$

We will be assuming an allowable deflection of Span/300 for beam elements with long term deflections based on dead load plus one third live load multiplied by a long term creep multiplier. The long term creep multiplier is 2 for seasoned timber and 3 for unseasoned timber.

6.4   Beams

Bending:

When timber beams bend the top half of the beam becomes in compression and the bottom half in tension, with maximum compression and tension stresses occurring at the top and the bottom. The stress in the centre of the beam is zero.

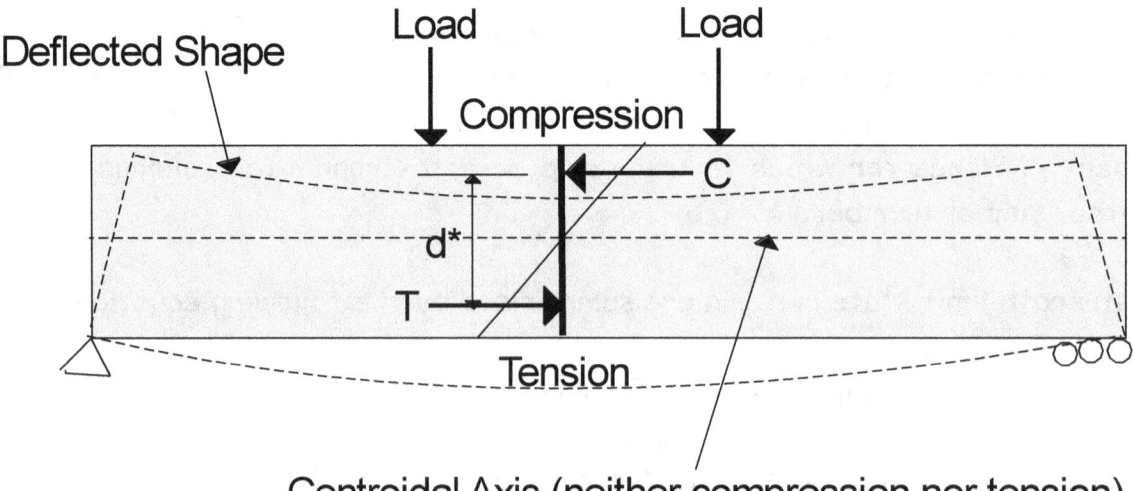

The distance between the centroid of the compression area and the centroid of the tension area (d^* in the diagram above) is equal to 2/3 times the depth of the beam (2/3×d). The compression force is equal to the tension force and is equal to 0.5×b×(d/2)× f'b where

f'$_b$ = the failure stress of the timber in bending

The tension force and the compression force form a "couple" to resist the applied moment equal to their force times the distance between their centroids or

Moment Capacity = 0.5×b×(d/2)× f'b×2/3×d = (b×d^2/6)×f'b

(b×d^2/6) is the section modulus of the section = Z

Hence Moment Capacity = Z×f'b

Applying a φ Factor we can express the member capacity φM_b as

$$\phi M_b \text{ (kN.m)} = \phi \times k_1 \times f'_b \times Z/10^6$$

where

Z = Section Modulus of the beam = b×d^2/6 (b and d in mm).

f'$_b$ = Characteristic Bending Strength of timber ≈ 3 Times the stress grade. eg if timber is F7 then f'$_b$ = 21 MPa

k_1 is a "duration of load" factor which takes into account the fact that the longer the element is loaded the weaker it becomes

k_1 = 1 for instantaneous loads

        =0.8 for combinations of dead plus live loads

        =0.57 for dead loads and permanent portions of live loads

ϕ = Capacity Reduction Factor =0.65 (Primary Structural Element) = 0.8 otherwise

To be Safe ϕM_b must be greater than M*

Shear:

The shear strength of a timber beam is dependent on the material properties and on the amount of "effective" material in the "centre" of the beam (A_s). For timber beams this is equal to two-thirds of the cross-sectional area (Note that for steel I beams the effective shear area is the area of the web).

$$\phi V = \phi \times k_1 \times f's \times A_s / 10^3 \quad \text{(kN)}$$

Where

A_s = 2/3×Cross Sectional Area of Beam = 2/3×b×d (b,d in mm)

f's = Characteristic Shear Strength

≈ [(0.21 × Stress Grade) + 0.73] MPa

To be Safe ϕV must be greater than V*

Deflection (δ):

Deflection of timber beams is calculated using the formulae given in Section 4.9 (eg for distributed loads on continuous beams $\delta = \frac{w \times L^4}{185 \times E \times I}$.).

E = 7,900 MPa (N/mm^2) F7 Timber (Eg Oregon)
E= 9,100 MPa F8 Timber
E = 10,500 MPa F11 Timber (Eg Unidentified Hardwoods)
E = 12,000 MPa F14 Timber
E = 14,000 MPa F17 Timber

I = Moment of Inertia of Beam (measure of the stiffness of the section geometry)
= $b \times d^3/12$ (mm^4) (b,d in mm)

Note long term deflection multiplier (j_2) = 3 for unseasoned timber and 2 for seasoned timber. For distributed loads the long term deflection is equal to the deflection due to (dead load plus one third of the live load) multiplied by j_2.

For point live loads the long term deflection is equal to j_2 times the dead load deflection plus the deflection due to the point live load.

For short term deflection calculations j_2 = 1

To be Safe δ must be less than δ (allowable)

δ (allowable) = L/300 (L in mm) unless specified otherwise

Example 1: A man stands in the middle of a 250 by 38 F8 Oregon Plank spanning 2.5 metres. Is the plank adequate, assuming that we wish to limit the short term deflection to 50 mm for some reason? Assume k_1 = 1 (Instantaneous load) and full live load for deflection.

Ignoring the dead load of the beam (which is small) the load on the beam is a live load with a load factor of 1.5. Assuming the man is 100kg (1kN) then

P* = 1.5 × 1 = 1.5 kN
Bending moment (M*) = PL/4 = 1.5 × 2.5 /4 = 0.94 kN.m
Shear Force (V*) = P/2 = 1.5/2 = 0.75 kN
Bending Check:
Z = 250 × 38^2/6 = 60,167 mm^3
f'b ≈ 3 × 8 = 24 MPa (assuming F8 Oregon)
φ = 0.65: k_1 = 1 (Instantaneous load)

Then φM_b = 0.65 × 1 × 24 × 60,167 /10^6 = 0.94 kN.m >= 0.94 kN.m Therefore O.K.

Shear Check:

f's ≈ (0.21 ×8) + 0.73 = 2.4 MPa A_s = 2/3 × 38 × 250 = 6,333 mm^2

then

φV = 0.65 × 1.0 × 2.4 × 6,333 /10^3 = 9.9 kN > 0.75 kN Therefore O K

Deflection Check:

E = 9,100 MPa   I = 1/12 × 250 × 38^3 = 1.143 × 10^6 mm^4

Note that because we are after short term deflection j_2 =1 and because the load is a point live load the full value is used rather than 70% which would be used for a distributed live load. Also note that the load used is unfactored.

δ = PL^3/48EI = 1000 × 2500^3/(48 × 9100 × 1.143 × 10^6) = 31 mm < 50

Therefore Acceptable

Example 2: A 50mm wide by 150 mm deep F14 unseasoned hardwood beam spans 3 metres and supports a design live load of 1.3 kN/m. Check strength and long term deflection assuming allowable long term deflection is span/300.

Dead load of plank = 0.05×0.15×11 = 0.09 kN/m

(Note 11 is unit weight of unseas.hwd)

Factored load for strength = w* = 1.2×0.09+1.5×1.3 = 2.06 kN/m

Long term load for deflection = w = 0.09 + 0.33×1.3 = 0.52 kN/m

Bending Check:

Bending moment (M*) = w×L^2/8 = 2.06×3^2/8 = 2.32 kN.m

Z = 50 × 150^2/6 = 187,500 mm^3

f'b ≈ 3 × 14 = 42 MPa

φ = 0.65: k_1 = 0.8 (dead load plus live load)

Then φM_b = 0.65 ×0.8 × 42 × 187,500 /10^6 = 4.10 kN.m >= 2.32 kN.m Therefore O.K.

Shear Check:

Shear Force (V*) = w*L/2 = 2.06×3/2 = 3.09 kN

f's ≈ (0.21 ×14) + 0.73 = 3.67 MPa A_s = 2/3 × 50 × 150= 5,000 mm^2

then

φV = 0.65 × 0.8 × 3.67 × 5,000 /10^3 = 9.54 kN > 3.09 kN Therefore O K

Deflection Check:

E = 12,000 MPa   I = 1/12 × 50 × 150^3 = 14.06 × 10^6 mm^4

δ = (5wL^3/384EI) ×j_2 = [5×0.52×3000^4/(384 × 12,000 × 14.06 × 10^6)]×3 = 9.8 mm

δ(allowable) =3000/300 =10 mm >9.8 Therefore acceptable.

(Note that for timber beams long term deflection is often the governing criteria)

6.5 Design of Floor Joist using AS1684 Supplement

The supplements to Australian Standard AS1684.2 Timber Framing Code give tables which are used to directly determine timber sizes. These supplementary tables are based on timber design principles previously dealt with. For beam type elements this essentially involves checking for bending, shear and deflection. Table 6.1 is a synthesis of these tables for the case of floor joists.

Consider seasoned F7 joists spaced at 450 and spanning 3 metres (single span). Using Table 6.1 we see that 190 by 35 joists will do. If we had continuous spans of 3 metres each then 170 by 35 joists would do.

Table 6.1 Floor Joist Spans Based on AS 1684.2

Joist Size	Seasoned Softwood Stress Grade F7	Seasoned Softwood Stress Grade F7
Joist Spacing 450 mm	Simply Supported	Continuous
120×45	2000	2300
140×35	2100	2500
140×45	2300	2800
170×35	2600	3200
170×45	2900	3500
190×35	3000	3600
190×45	3300	4000
240×35	4000	4900
240×45	4400	5300
290×45	5200	6100
Joist Spacing 600 mm	Simply Supported	Continuous
120×45	1800	2100
140×35	2000	2300
140×45	2200	2500
170×35	2400	2800
170×45	2700	3100
190×35	2800	3200
190×45	3000	3500
240×35	3600	4200
240×45	3900	4600
290×45	4800	5600

6.6   Design of Floor Joist using "Smartframe" program

The same joist can be designed using the program "Smartframe" (Available free from Tilling Timber – www.tilling.com.au).

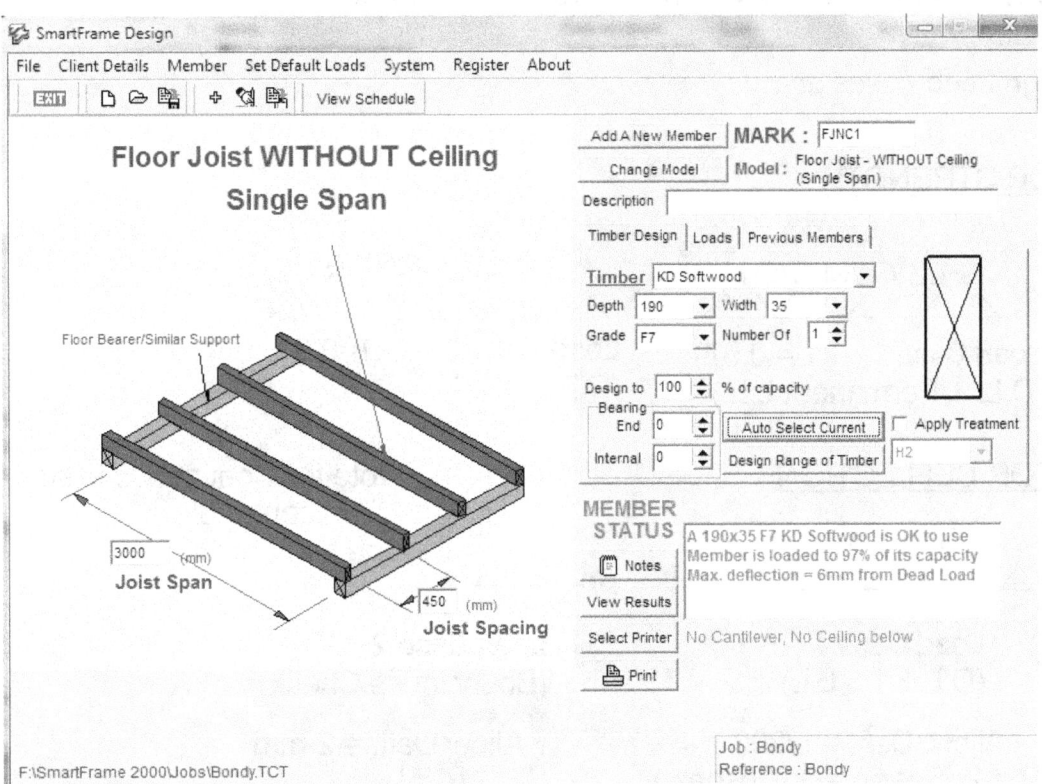

Figure 6.1   Output from "SmartFrame " program

6.7   Design of Floor Joist from First Principles

AS 1684.1 gives the design criteria by which timber sizes are worked out for the Supplements to the Timber Framing Code AS1684.2.

AS 1684.1 sets out the design basis for floor joists as follows. The design of floor joists will be used as an indication of how the sizes were determined in AS1684.2. **We will not do shear checks**

1. Dead load of floor (G_1) assumed to be 40 kg/m^2 + DL of joist, say take total of 50 kg/m^2,i.e 0.5 kPa.
2. Permanent live load (Q_1) assumed to be 0.5 kPa, i.e 0.33 times total live load
3. Total live load (Q_2) assumed to be 1.5 kPa.
4. New factor k_9 not previously covered. k_9 is a factor for <u>strength</u> sharing of joists
   $k_9 = 1+0.26\times(1-2\times S/L)$ where S = Joist Spacing and L = Joist Span.

NOTE k_9 should always be greater than 1.

5. New point load Q_7 introduced with deflection limit of 2 mm . This is a check on bounciness of floor due to live load. Formula complicated but we will assume $Q_7 = 0.5$ kN
6. Factor j_2 introduced for deflections
   Long term Deflection = $j_2 \times$Calculated Short Term Deflection
   Deflection limits based on long term deflection

Design Load cases are

FOR STRENGTH

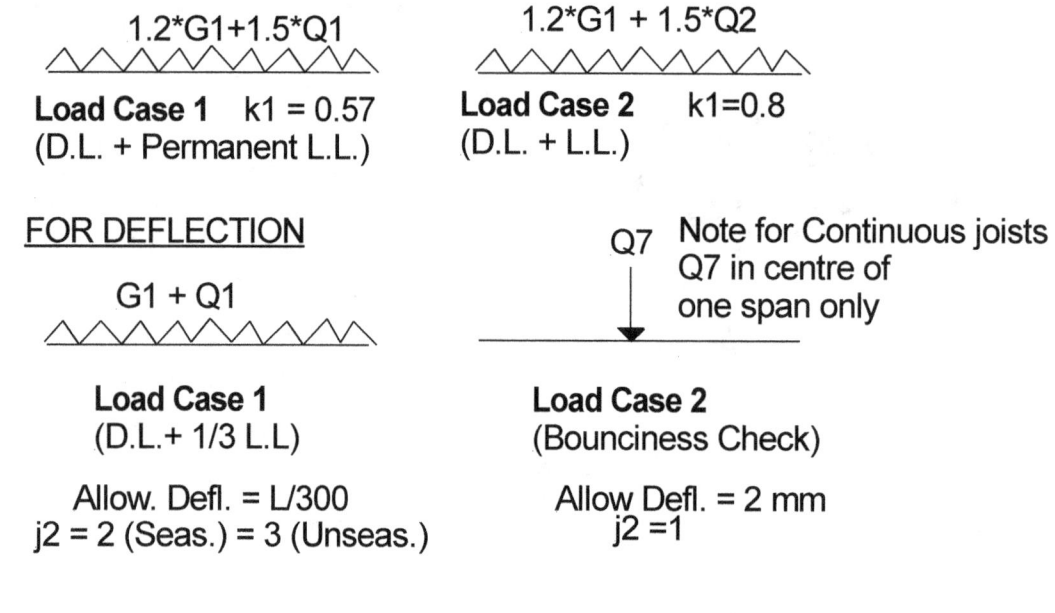

$$1.2*G1+1.5*Q1$$

Load Case 1   k1 = 0.57
(D.L. + Permanent L.L.)

$$1.2*G1 + 1.5*Q2$$

Load Case 2   k1=0.8
(D.L. + L.L.)

FOR DEFLECTION

$$G1 + Q1$$

Load Case 1
(D.L.+ 1/3 L.L)

Allow. Defl. = L/300
j2 = 2 (Seas.) = 3 (Unseas.)

Q7 Note for Continuous joists
Q7 in centre of
one span only

Load Case 2
(Bounciness Check)

Allow Defl. = 2 mm
j2 =1

(FOR BM's etc see Section 4.9)

Joist Example

Consider 190 × 35 Seasoned F7 joists spaced at 450 and spanning 3 metres (single span) – The checks are

Loads and Factors

$G_1 = 0.5 \times .45 = 0.23$ kN/m
$Q_1 = 0.5 \times 0.45 = 0.23$ kN/m
$Q_2 = 1.5 \times 0.45 = 0.68$ kN/m
$k_9 = 1 + 0.26 \times (1 - 2 \times 450/3000) = 1.18$
$Q_7 = 0.5$ kN

Bending Check

$$1.2*0.23+1.5*0.23 = 0.62 \text{ kN/m} \qquad 1.2*0.23 + 1.5*0.68 = 1.30 \text{ kN/m}$$

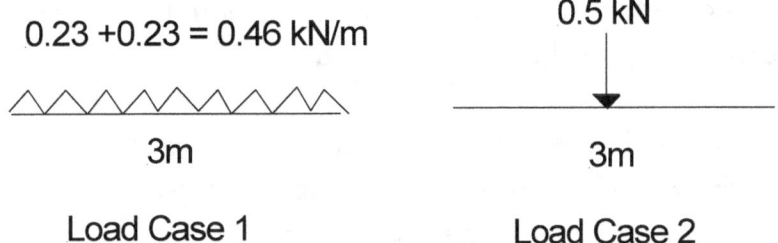

3m   3m

Load Case 1   k1 = 0.57   Load Case 2   k1=0.8

Case 1

$$\phi M_b = \phi \times k_1 \times k_9 \times f'_b \times Z/10^6 = 0.8 \times 0.57 \times 1.18 \times (3 \times 7) \times (35 \times 190^2/6)/10^6$$
$$= 2.38 \text{ kN.m}$$
$$M^* = 0.62 \times 3^2/8 = 0.70 \text{ kN.m} < 2.38 \text{ Therefore OK}$$

Case 2

$$\phi M_b = \phi \times k_1 \times k_9 \times f'_b \times Z = 0.8 \times 0.80 \times 1.18 \times (3 \times 7) \times (35 \times 190^2/6)/10^6 = 3.34 \text{ kN.m}$$
$$M^* = 1.30 \times 3^2/8 = 1.46 \text{ kN.m} < 3.34 \text{ Therefore OK}$$

Deflection Check

$$0.23 + 0.23 = 0.46 \text{ kN/m} \qquad\qquad 0.5 \text{ kN}$$

3m   3m

Load Case 1   Load Case 2

Case 1

$$\delta = \frac{5 \times w \times L^4}{384 \times E \times I} \times j_2 = \frac{5 \times 0.46 \times 3000^4}{384 \times 7900 \times 20 \times 10^6} \times 2 = 6.1 \text{ mm}$$

$\delta_{allowable}$ = Min 3000/300=10

Since 10 > 6.1 deflection OK

Case 2

$$\delta = \frac{P \times L^3}{48 \times E \times I} \times j_2 = \frac{500 \times 3000^3}{48 \times 7900 \times 20 \times 10^6} \times 1 = 1.8 \text{ mm}$$

$\delta_{allowable}$ = 2 mm > 1.8 Therefore OK

6.8   Tension Members

Tension members occur mainly in trusses but can also occur in a wide variety of situations.   The principal concern with tension members is that the tensile force in the member N* does not exceed the failure load of the element ϕN_t .
(Note that if a bar is pulled at each end with a factored force of 100kN then N* = 100 kN not 200 kN as is often incorrectly assumed by students.)
The failure load of timber loaded parallel to the grain is simply the net cross sectional area of the element (i.e. minus any holes) times the tensile failure stress of the material, modified by the capacity reduction factor (ϕ) and a duration of load factor (k_1).

$$\phi N_t = \phi \times k_1 \times A_n \times F_t \ /1000$$

ϕN_t = Capacity of Member in Axial Tension (kN)

A_n = Net Cross Sectional Area of Member (mm^2) (Allow an extra 2 mm on the bolt diameter for bolt holes.)
k_1 = Load duration Factor

For timber members the tensile failure stress (F_t) of various common timbers in Australia loaded parallel to the grain may be assumed to be 1.8 times the stress grade, so for example F_t for average F11 hardwood would be 1.8×11 = 19.8 MPa, for F7 mechantable grade Oregon 1.8×7 = 12.6 MPa.

In the majority of cases the capacity of tensile members is limited by the connections, whether they be nails or bolts. Table 6.2 gives an idea of the capacities of various connections <u>loaded parallel to the grain</u>

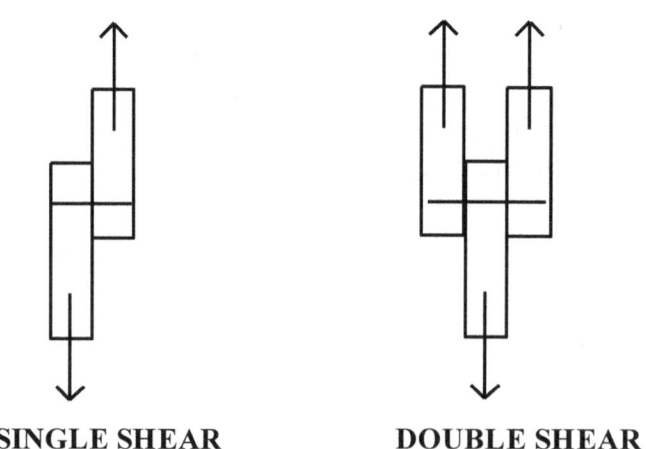

SINGLE SHEAR   **DOUBLE SHEAR**

Table 6.2   Shear Capacities of Joints

Capacity of Connector (kN) assuming ϕ = 0.65 ,k_1 =1 and an effective timber thickness of 50 mm			
Joint	Joint Type	Seasoned Oregon	Unseasoned Hardwood
1/ 3.15 mm Nail	Single Shear	0.5	0.65
1/ 4.5 mm Nail	Single Shear	0.8	1.0
1/M12 Bolt	Single Shear	7.0	6.6
	Double Shear	14.0	13.2
1/M24 Bolt	Single Shear	11.0	13.8
	Double Shear	22.0	27.6

Example:

An F7 seasoned oregon tension member is required to resist a live load (k_1=0.8) tension force of 5 kN in a single shear joint. What size bolt is required and will a 50 by 50 size do ?

Assuming a load factor of 1.5

N* = 1.5 × 5 = 7.5 kN

< (0.8×11.0) = 8.8 kN therefore M24 OK (Note k1 =0.8 hence ×0.8)

A_n = (50×50)– (26×50) = 1200 mm^2

ϕN_t = 0.65 ×0.8× 1200 ×(1.8×7) /1000 = 7.86 kN

Since 7.86 kN is greater than 7.5 kN 50 by 50 size OK

6.9   Compression Members

Pure compression members occur in timber structures in the form of compressive struts in trusses and in timber posts.

Adopting a capacity reduction factor ϕ as for tension members we can write the equation as

ϕN_c = Compression capacity of column (kN)
   = $\phi \times k_1 \times k_{12} \times f'c \times A_c$ / 1000 (kN) where

ϕ = Capacity Reduction Factor as above

k_1 = Duration of Load factor as above

k_{12} = Factor for effect of Column Length
      = $1.4 \times e^{(-0.026 \times l/r)}$
      = 1.0 if l/r <17

The capacity of **short** compression members is limited by the crushing capacity of the material in compression F'_c and the cross-sectional area of the member. For **long** compression members (where the height is greater than about 10 times the least dimension) the resistance of the compression member to buckling will determine its compression capacity. <u>This is where the k_{12} factor comes in.</u>

"r" is a Section Property called the "Radius of Gyration" (see Appendix C) and measures the ease of which a section shape buckles. In most cases r is taken as r_{yy} if buckling (bending) can occur about the weak yy axis.

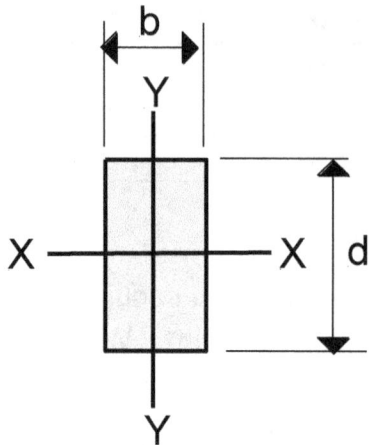

$r_{yy} = 0.29 \times b$ for rectangular sections (b = Min Dim)
l = Column or strut length (Ignoring end conditions)

f'c = Characteristic Compressive Strength of timber
$\approx 2.25 \times$ Stress Grade (Mpa)
A_c = b × d (mm^2)

To be satisfactory N* (Factored Compressive Load in Strut) must be less than ϕN

Example:
A 100 by 100 F7 timber column in a house is required to carry a permanent dead load of 7 kN. If it is 2.5 metres high is it O.K.

N* = 1.2×7 = 8.4 kN
ϕ =0.65 (Primary Structural Element)
Assume pinned ends therefore l = 2500 mm
r = 0.29 × 100 = 29 mm
$k_{12} = 1.4 \times e^{(-0.026 \times 2500/29)} = 0.15$ $k_1 = 0.57$ (Permanent)
$\phi N = 0.65 \times 0.57 \times 0.15 \times (2.25 \times 7) \times (100 \times 100)$
= 8,775 N = 8.8 kN > 8.4 kN
Therefore section is adequate.

EXERCISES:

1. A 200 wide by 300 deep F14 seasoned hardwood beam is 10 metres long and carries a dead load of 0.5 kN/m in addition to its own dead load and a point live load of 2 kN in the centre. Check bending, shear and long term deflection assuming a limit of 60 mm. Assume $k_1 = 0.8$ and $\phi = 0.65$. Note for maximum deflection use $j_2 = 2$ for dead load and $j_2 = 1$ for point live load.

2. Using Table 6.2 choose a suitable single span seasoned F7 floor joist to span 2.4 metres if the joists are at 600 mm centres. Check Load cases 1 and 2 for strength and deflection.

3. Using Table 6.2 choose a suitable <u>continuous</u> span seasoned F7 floor joist to span 4 metres if the joists are at 450 mm centres. Assume two spans for calculating moment and deflection (Section 4.9). Check Load cases 1 and 2 for strength and deflection.

4. The bottom chord of a truss is a 75 by 50 seasoned F8. It carries a factored dead load tension force of 18 kN. Assuming nailed joints(i.e. no reduction in area) is it safe ?. Take $k_1 = 0.57$ and $\phi = 0.65$.

5. A 150 by 75 F14 column is 3 metres long and carries a live load of 4.5 kN. Assuming $k_1 = 0.8$ check whether it is safe

7. DESIGN OF SIMPLE STEEL AND CONCRETE ELEMENTS

7.1 Steel Members

7.1.1 Bending

The equation for bending moment capacity of a steel beam <u>with continuous support of the compression flanges</u> is basically the same as that for timber beams except there is no duration of load factor (k_1) as steel properties do not change with time.

$$\phi \, M_b = \phi \times f_y \times Z_{eff} / 10^6 \quad \text{(kN.m)}$$

where

$\phi = 0.9$

f_y = Yield Stress of steel = 250 MPa generally

$Z_{eff} = 1.15 \times Z$ where Z= Section Modulus of Beam (See Appendix C6). The 1.15 factor takes into account the yielding of steel beams at failure.

Where the <u>compression flanges of steel beams (usually top flange) are not continuously supported</u> they can buckle sideways under the applied loads, similar to how columns buckle sideways under compression loads. The buckling moment of such a beam is dependent on the ratio of the effective length (l_e) to the radius of gyration of the section (r_{yy}) and on the shape of the bending moment diagram. The former is reflected in a k_{12} factor (similar to the k_{12} factor in timber columns) whilst the latter is reflected in a a_m (alpha m) factor. For these beams the moment capacity is

$$\phi \, M_b = \phi \times k_{12} \times a_m \times f_y \times Z_{eff} / 10^6 \quad \text{(kN.m) where}$$

$k_{12} = 1.22 \times e^{(-0.0065 \times l_e / r_{yy})}$
$\quad\quad$ =1 for $l_e/r_{yy} < 30$

l_e = effective length of the segment
$\quad \approx 1.1 \times$ Span of the beam for beams with no intermediate supports
$\quad \approx 1.1 \times$ Distance between intermediate points of lateral restraint to compression
$\quad\quad$ flange (a lateral restraint occurs when the compression flange is prevented
$\quad\quad$ from moving sideways by another member at right angles to the beam)

r = Radius of gyration of the beam about y-y axis = r_{yy} (Calculated as $(I_{yy}/A)^{0.5}$ or given in Tables -See Appendix C6)

Note that if beam is loaded on the top flange and the top flange can rotate increase l_e by 40% This is because a load on the top flange tends to cause a beam to rotate and buckle easier.

a_m ≈ 1.13 for simply supported beams with no intermediate restraints
    ≈ 1.33 for simply supported beams with central restraint
    ≈ 1.0 for beams with intermediate supports for positive BM capacity
    ≈ 2.25 for continuous beams with no intermediate restraints
    ≈ 1.75 for continuous beams with intermediate restraints for negative BM capacity
    = 1 when k_{12} = 1

Remember that when calculating M* always use the actual length "L" and not the effective length "l_e".

The formula given above for $\phi\,M_b$ is an approximation only but gives very good results for usual beam sections (See Figure 7.1)

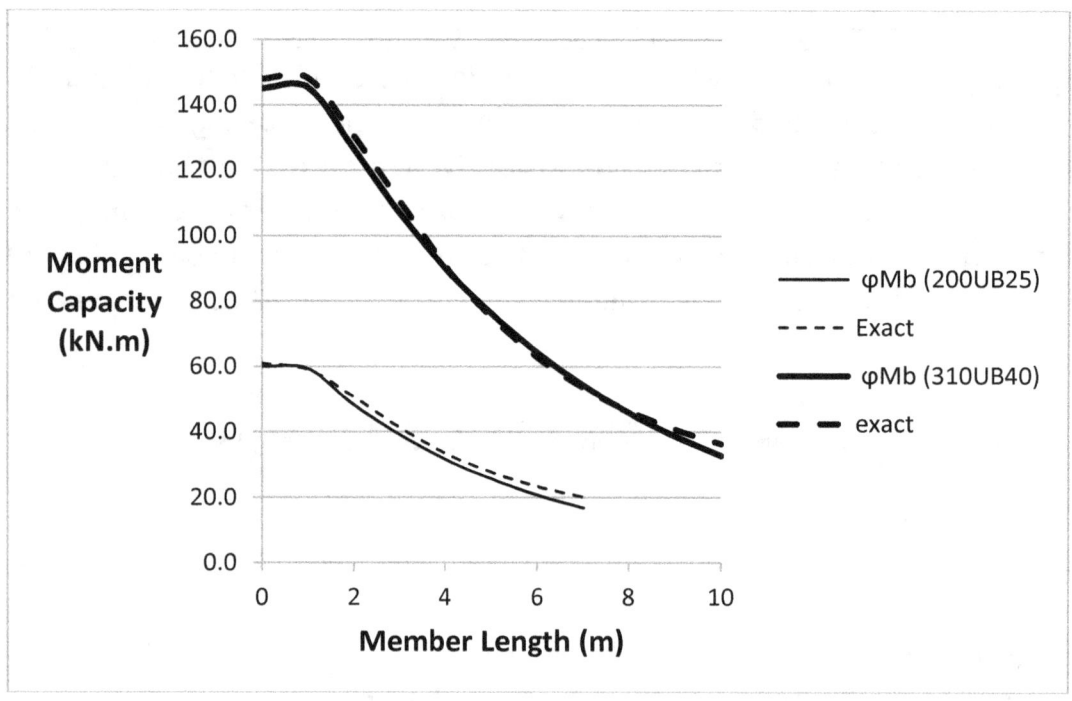

Figure 7.1 Comparison of $\phi\,M_b$ Formula above with Exact Solutions

Figure 7.2 illustrates the values of l_e (effective length) and a_m (alpha m) in the case of uniformly loaded beams.

86

X = Lateral Restraint

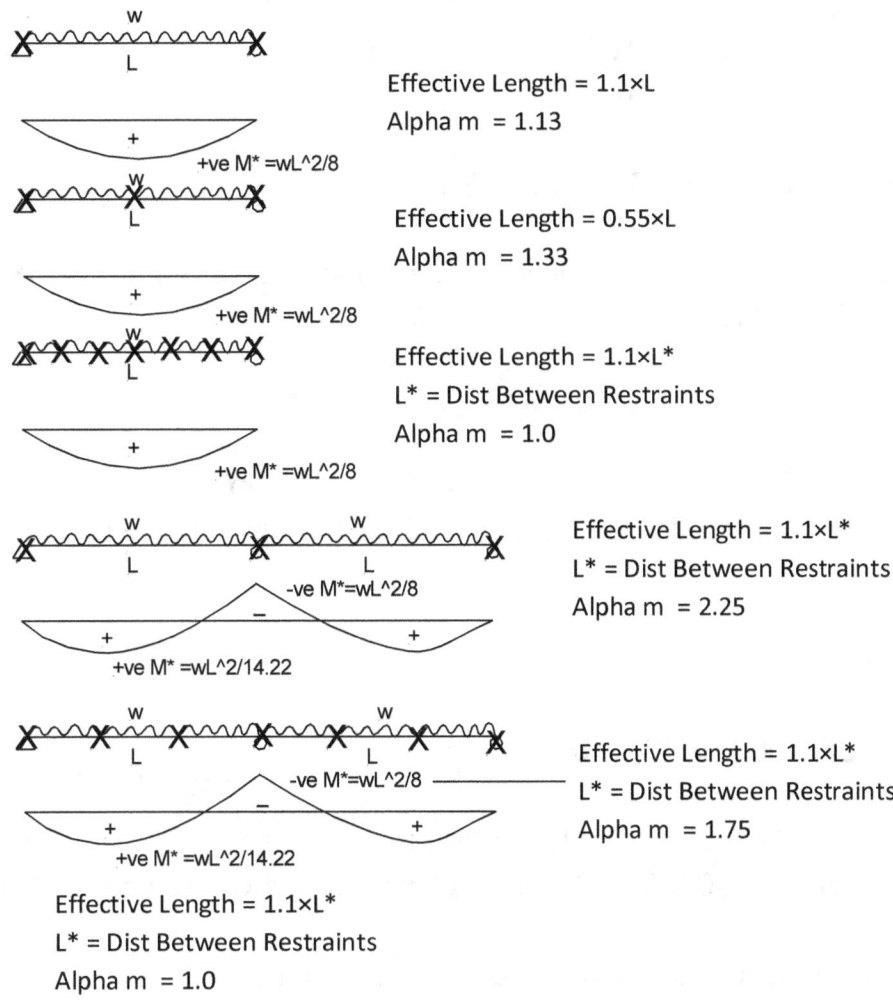

Effective Length = 1.1×L
Alpha m = 1.13

Effective Length = 0.55×L
Alpha m = 1.33

Effective Length = 1.1×L*
L* = Dist Between Restraints
Alpha m = 1.0

Effective Length = 1.1×L*
L* = Dist Between Restraints
Alpha m = 2.25

Effective Length = 1.1×L*
L* = Dist Between Restraints
Alpha m = 1.75

Effective Length = 1.1×L*
L* = Dist Between Restraints
Alpha m = 1.0

Figure 7.2 Values of l_e (effective length) and α_m (alpha m)

7.1.2 Shear

Shear is not generally a problem unless the loads are very large and the spans are short.

7.1.3 Deflection

E = 200,000 MPa (2×10^5 MPa), I from tables (eg Appendix C6). Allowable deflection = Span/300

Load to be used is dead load plus 70% of the live load in the case of uniformly distributed loads and 100% of the point loads. <u>Where the beam size is not given assume dead load in kN/m is equal to 0.05 times the span in metres.</u>

For uniformly loaded beams putting the allowable deflection equal to span/300 gives

For simply supported beams $I_{required}$ = 0.02× w×L^3 ×10^6 mm^4
For continuous beams $I_{required}$ = 0.008× w×L^3 ×10^6 mm^4
Where w is in kN/m and L is in metres

Example 1: Design of Simply Supported Beam

Determine the size of a simply supported beam to span a distance of 6 metres and carry a live load of 5 kN /m.

Estimate of dead load = 0.05×6 =0.3 kN/m

In this case $I_{required}$ = 0.02×(0.3+0.7×5) × 6^3 = 16.4 × 10^6 mm^4
from Appendix C6 choose a 200UB 25 I_{xx} = 23.6×10^6 mm^4, Z_{xx} = 232,000 mm^3 and r_{yy} = 31 mm

<u>Check deflection</u>

Dead Load = 25×10/1000 = 0. 25 kN/m
w = 0.25 + 0.7×5 = 3.75 kN/m
δ = 5×3.75×6000^4/(384×2×10^5×23.6×10^6) = 14 mm
Allowable Deflection = 6000/300 = 20 mm >14 therefore OK for deflection

Next <u>check strength</u> assuming no intermediate restraints

Factored Load = 1.2×0.25 + 1.5×5 = 7.8 kN/m
M* = 7.8 × 6^2/8 = 35.1 kN.m

l_e = 1.1×6 = 6.6 m
Z_{eff} = 1.15×232,000 =266,800 mm^3
k_{12} = 1.22 × e$^{(-0.0065×6600/31)}$ = 0.306
φ M_b = φ × k_{12} × a_m × f_y × Z_{eff} / 10^6 (kN.m)
        = 0.9 × 0.306 × 1.13 × 250 ×266,800/10^6 = 20.8 kN.m
< 35.1 (M*) Therefore 200 UB 25 NOT OK for strength

Try 250 UB 31 (will be OK for deflection so just check strength)

From Appendix I_{xx} = 44.4×10^6 mm^4, Z_{xx} = 353,000 mm^3 and r_{yy} = 33.5 mm

Z_{eff} = 1.15×353,000 =406,000 mm^3
k_{12} = 1.22 × e$^{(-0.0065×6600/33.5)}$ = 0.34
φ M_b = φ × k_{12} × a_m × f_y × Z_{eff} / 10^6 (kN.m)

$= 0.9 \times 0.34 \times 1.13 \times 250 \times 406{,}000/10^6 = 35.1$ kN.m

$= 35.1$ (M*) Therefore 250 UB 31 OK for strength and deflection

Alternatively if we put in a lateral restraint to the top flange at the middle of the 200 UB 25 beam

$l_e = 0.55 \times 6 = 3.3$ m and

$k_{12} = 1.22 \times e^{(-0.0065 \times 3300/31)} = 0.61$ and

$\phi\, M_b = 0.9 \times 0.61 \times 1.33 \times 250 \times 266{,}800/10^6 = 48.7$ kN.m

> 35.1 (M*) Therefore 200 UB 25 OK with lateral restraint to top flange in centre.

Example 2: Check of Continuous Beam

A 310 UB 40 spans 2 spans of 7 metres and carries a dead load of 25 kN/m in addition to its own dead weight. It has intermediate lateral restraints at the third points. Is it safe?

From Section C6 $I_{xx} = 85.2 \times 10^6$ mm4 $Z_{xx} = 561{,}000$ mm^3 and $r_{yy} = 38.5$ mm

Check Deflection

Dead load = 25 + 0.4 = 25.4 kN/m

From Section 4.9

$\delta = 25.4 \times 7000^4/(185 \times 2 \times 10^5 \times 85.2 \times 10^6) = 20$ mm

Allowable deflection = 7000/300 = 23 mm >20 Therefore OK for deflection

Check Strength

Factored Load = 1.2 × 25.4 = 30.5 kN/m

Positive M* = $30.5 \times 7^2/14.22 = 105$ kN.m

$l_e = 1.1 \times (7/3) = 2.57$ m

$Z_{eff} = 1.15 \times 561{,}000 = 617{,}100$ mm^3

$k_{12} = 1.22 \times e^{(-0.0065 \times 2570/38.5)} = 0.79$

$\phi\, M_b = \phi \times k_{12} \times a_m \times f_y \times Z_{eff} / 10^6$ (kN.m)

$= 0.9 \times 0.79 \times 1.0 \times 250 \times 617{,}100/10^6 = 110$ kN.m

>105 (+veM*) Therefore OK

Negative M* = $30.5 \times 7^2/8 = 187$ kN.m

$\phi\, M_b = \phi \times k_{12} \times a_m \times f_y \times Z_{eff} / 10^6$ (kN.m)

$= 0.9 \times 0.79 \times 1.75 \times 250 \times 617{,}100/10^6 = 192$ kN.m

>187 kN.m (-veM*) Therefore OK

7.1.4 Approximations of Section Properties

Appendix C gives a detailed method on how to calculate section properties for any section and Section C6 gives values of I_{xx}, Z_{xx} and r_{yy} for universal beams (UB's).

Where sections are symmetrical such as universal beams, i.e their centroidal axis is in the middle it is relatively easy to calculate section properties. Consider the case of the following approximation of a 180 UB 18. We are usually only interested in the moment of inertia about the x-x axis (I_{xx}), the section modulus about the x-x axis (Z_{xx}) and the radius of gyration about the weak y-y axis (r_{yy}).

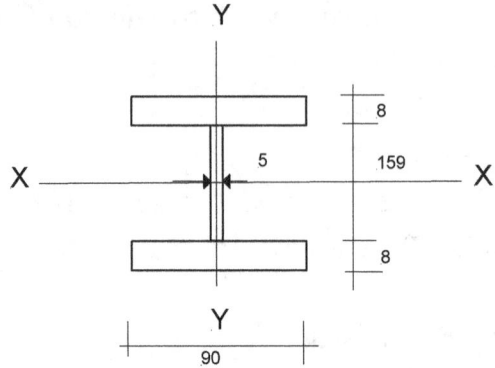

For properties about the X-X axis we need to find I_{xx} first.

I_{xx} = I_{xx} for the overall size – I_{xx} for the voids

$$= 90 \times 175^3/12 - 85 \times 159^3/12 = 11.72 \times 10^6 \text{ mm}^4$$

The exact value of 12.1×10^6 mm^4 (3% greater) given in Appendix C6 reflects the additional steel in the radius sections where the flange joins the web.

Z_{xx} is I_{xx} divided by the distance from the centre to the top of the beam

$Z_{xx} = 11.72 \times 10^6/87.5 = 134 \times 10^3$ mm^3 (Exact value 139×10^3 mm^3)

Similarly we can calculate

$I_{yy} = 2 \times 8 \times 90^3/12 + 159 \times 5^3/12 = 0.974 \times 10^6$ mm^4

The radius of gyration about the weak y-y axis (r_{yy}) = $(I_{yy}/A)^{0.5}$

$r_{yy} = [0.974 \times 10^6/(2 \times 90 \times 8 + 159 \times 5)]^{0.5} = 20.9$ mm (exact 20.6 mm)

Where the exact dimensions of symmetrical steel sections are not known the following approximations may be used.

For Flanged Sections $I_{xx} \approx 23 \times M \times D^2$ (mm^4)

Where M = Mass /m in kg , D is overall depth in mm

$Z_{xx} \approx 46 \times M \times D$

r_{yy} $\approx 2.15 \times \sqrt{D}$ for UB's with D>=200

          $\approx 1.5 \times \sqrt{D}$ for D< 200

          $\approx 0.25 \times D$ for UC's

          $\approx 0.39 \times B$ for RHS and SHS sections where B = min dim

          $\approx 0.35 \times$ Diameter for CHS's

For example for a 200UB 30 (Depth 200 mm and mass/m = 30kg)

$I_{xx} \approx 23 \times 30 \times 200^2 = 27.6 \times 10^6$ mm4 (Exact value 28.9×10^6 – error 4.5%))

$Z_{xx} \approx 46 \times 30 \times 200 = 276 \times 10^3$ mm3 (Exact value 279×10^3 – error 1.1 %)

$r_{yy} \approx 2.15 \times 200^{0.5} = 30.4$ mm (Exact value 31.8 mm – error 4.4%)

7.2 Reinforced Concrete Beams

For concrete beams the width is generally set based on architectural considerations and the depth must be enough to ensure deflection is within the allowable limit. As an initial guess for depth refer to Table 7.1 which relates depth to the span of the beam. This may need to be changed if the beam fails the strength or serviceability checks.

Table 7.1 Trial Span to Depth Ratios for Concrete Beams

Reinforced Concrete Beams	
Simply Supported	12.5
Continuous	15
Cantilever	6

7.2.1 Bending in Beams

In reinforced concrete (R.C.) members the steel is put in to resist any tension that occurs in the beam. For simply supported beams the tension is in the bottom of the beam but in continuous beams there is tension in the bottom between supports and tension in the top over supports. The depth to the <u>centre</u> of the tension resisting steel from the top of the compression face is called "d" to distinguish it from the overall depth of the beam (D). The cross sectional area of the main steel reinforcement is called "A_{ST}". This is what we need to calculate.

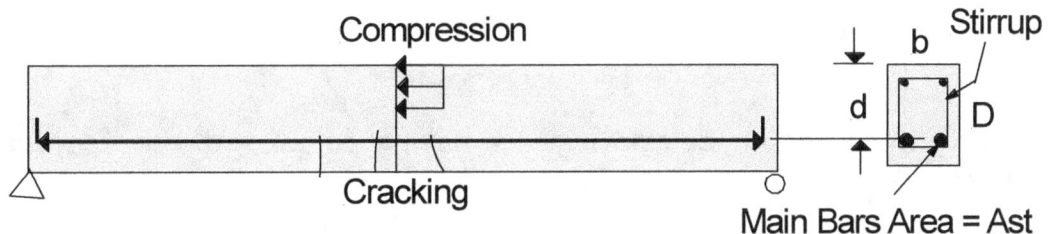

d = Overall beam depth (D) – Stirrup diameter (Assume 10 mm) – Cover (for required cover see Table 7.2)– 1/2×Diameter of Main Bars.

You must assume the diameter of the main bars that you are going to use before doing calculations. If you then have to use bigger bars to fit the steel in the width of the beam you will have to redo your calculations. The cross sectional area of one of the assumed bar diameters is given the symbol A_b. For simplicity A_b is assumed to be $\pi \times d^2/4$ where d is bar diameter. This is slightly different to the exact cross sectional area of bars due to N bars having deformed surfaces for better bond.

Steel Designations

Bars with a yield stress of 500 MPa are given the prefix "N" , eg N12 is 12 mm dia bar with a yield stress of 500 MPa. Available bar sizes are N12 (A_b=113mm^2), N16 A_b=201mm^2, N20 (A_b=314mm^2), N24 (A_b=452mm^2), N28 (A_b=616mm^2), N32 (A_b=804mm^2)

Trench Mesh is used sometimes in footings instead of bars. It consists of three or four main bars at 100 mm centres joined with light cross wires at 300 mm centres. Main bar sizes are 8mm, 11mm or 12mm. 8TM/3 denotes trench mesh with 3/8mm bars at 100 mm centres.

Moment Capacity

Using a similar analysis to that shown for timber beams in Section 6.4 it can be shown that

$$\text{At failure } \phi M_b \approx \frac{A_{ST} \times d}{2900} \text{ where}$$

A_{st} = Cross Section area of steel in tension zone and d is as defined above

Example 1: Consider the case of a 200 wide by 400 (span/12.5) deep beam spanning 5 metres and carrying a live load of 13 kN/m.

M* = (1.2×0.2×0.4×25+ 1.5×13)×5^2/8 = 68.4 kN.m

Assume that there are 2/N20 bars in the bottom with R10 stirrups and 50 mm cover to the stirrups then d = 400 -50(cover)-10(stirrup)-10(1/2 bar dia) = 330 mm and

$$\phi M_b \approx \frac{628 \times 330}{2900} = 71.5 \text{ kN.m} > M*$$

In general we are interested in calculating A_{st} having calculated the factored bending moment M*.

Making M* = ϕM_b gives

$$A_{st} = 2900 \times M*/d$$

In the above example if we put M* in the above equation we get

$$A_{st} \text{ Required} = 2900 \times 68.4/330 = 601 \text{ mm}^2$$
$$\text{Say 2/N20 bars } (A_{st} = 2 \times 314 = 628 \text{ mm}^2)$$

To get the required number of reinforcement bars of the chosen diameter divide the A_{st} calculated from the formula above by the area of one bar of the assumed diameter. For example if A_{st} turns out to be 601mm^2 and you have assumed N20 bars then the required number of bars is 601/314 = 1.9 and you would choose 2 N20 bars.

Note that the approximate formula for A_{st} given above is independent of concrete strength (F'c) but it is accurate enough for concrete strengths <= 40 MPa. It is also independent of beam width "b". However a check needs to be made that steel percentage (p = A_{st}/bd) is less than 0.01 (1%) to avoid compression failure in the concrete. In the above case p = 601 /(200×330) = 0.009 (0.9%) which is less than 1% so OK for compression failure. Note we use calculated value of A_{st} not what is there. If p>1% you will either need to make the beam deeper or put in top compression steel.

Example 2:

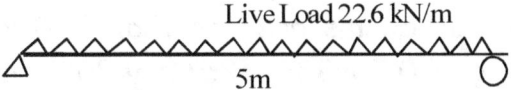

Live Load 22.6 kN/m

5m

Assume b= 300 mm, 45 Cover, F'c = 32MPa, R10 Stirrups and N20 bars
From Table 7.1 initial estimate of overall depth = 5000/12.5 = 400 mm.

Dead Load = 0.4×0.3×25 = 3 kN/m (25 is the unit weight of reinforced concrete)
Factored Load = 1.2×3 +1.5×22.6 = 37.5 kN/m

M* = $\dfrac{37.5 \times 5^2}{8} = 117 kN.m$

d = 400 −45−10−10 = 335 mm

Ast = $\dfrac{2900 \times 117}{335} = 1013 mm^2$

Therefore required number of N20 bars = 1013/314 = 3.2 say 4

Check Steel percentage $\leq$ 1%

p = $\dfrac{1013}{300 \times 335} = 0.01 \leq$ 1% OK

7.2.2 Shear:

In beams shear forces are highest near the supports and in these regions stirrups are more closely spaced.

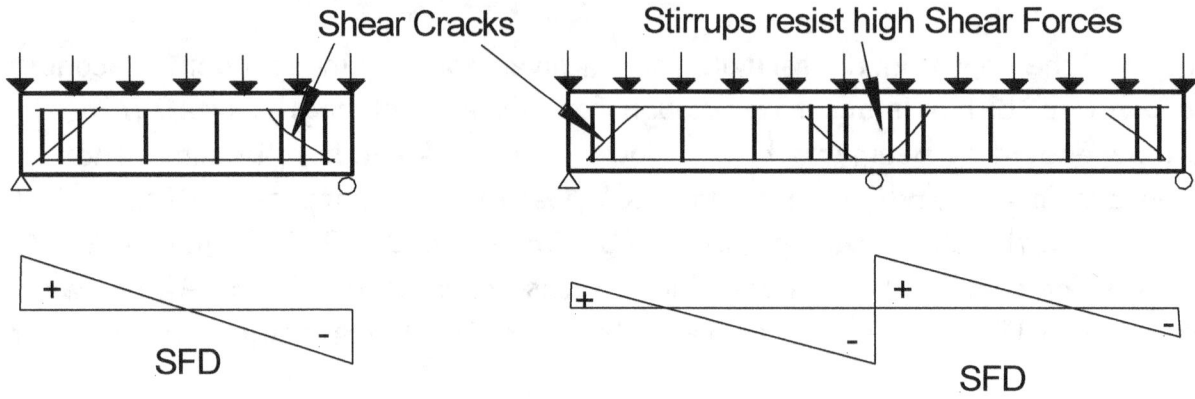

Where the shear stresses are not high R 10 stirrups at 0.75×d spacing will generally be conservative for beams greater than 300 wide. Thus in the above example R10 stirrup spacing of 0.75×335 = 251 say 250 mm would probably be OK. In slabs shear is rarely a problem except around points of concentrated loads around columns (punching shear)

7.2.3 Bending in Slabs:

Same as for beams except no stirrups and

$$Ast = 2650×M*/d$$

Steel Designations   Mesh SL - equal diameter bars at equal spacing in both directions (eg SL 72 is 7mm diameter bars at 200mm centres both ways) or RL – thicker bars one way at typically 100 centres with 8mm cross bars at 200 mm centres (eg RL 1218 has 12 mm bars at 100 mm centres in one direction and 8 mm bars at 200 mm centres in the other direction)
Bars as for beams

Limit on p (A_{ST}/bd) of 0.005 (0.5%). Note that for slabs we consider a nominal metre wide strip and therefore b = 1000 mm

7.2.4   Deflection of Reinforced Concrete Beams and Slabs

Deflection of reinforced concrete beams and slabs is based on the deflection formula given in Section 5.9 with a load equal to the dead load plus one third of the live load. The permanent portion of the live load is deemed to be one third of the total live load.

Allowable deflection is span/300 for visual appearance

I ≈ 0.045bd³ for beams and 0.033bd³ for slabs. Reinforced concrete beams (and slabs) always crack in the tension zone near the steel and this reduces their moment of inertia to about 55% of the uncracked moment of inertia ($0.083bd^3 = bd^3/12$) in the case of beams and 40% in the case of slabs.

Modulus of elasticity $E = 5055 \times F'^{0.5}_c$ where F'_c is the concrete strength.

long term creep factor = $[3-1.2 \times A_{sc}/A_{st}]$ where A_{sc} is the <u>actual</u> cross sectional area of steel in the compressive zone of the beam (top for simply supported members) and A_{st} is the <u>actual</u> (not calculated) cross sectional area of steel in the tension zone.

For Example 2 above, assuming 2/N16 in the top (A_{sc}/A_{st} = $(2 \times 201)/(4 \times 314)$ = 32%) and long term creep factor = $3 - 1.2 \times 0.32 = 2.62$
$I = 0.045 \times 300 \times 335^3 = 5.07 \times 10^8$ mm^4
$E = 5055 \times 32^{0.5} = 28,595$ MPa
Load = $3 + 0.33 \times 22.6 = 10.46$ kN/m
Long Term Deflection is
$$\delta = \frac{5}{384} \times \frac{10.46 \times 5000^4}{28,595 \times 5.07E8} \times 2.62 = 15.4 \text{ mm}$$

Allowable deflection = 5000/300 = 16.7 mm > 15.4 so the Beam OK for deflection

Note that in this case the short term load ($3 + 0.7 \times 22.6 = 18.8$ kN/m) divided by the creep factor of 2.62 = 7.2 kN/m which is less than load for the long term deflection (10.46 kN/m) and therefore long term deflection governs.

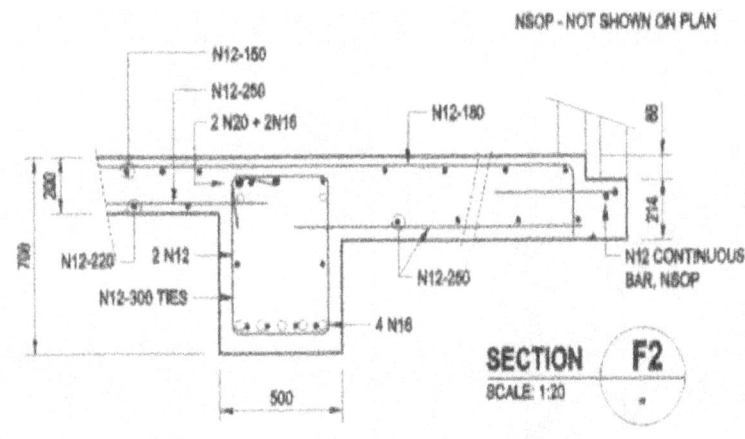

Figure 7.3 Typical Beam Cross Section

7.2.5 Durability Considerations

Durability in reinforced concrete relates to the protection of the reinforcing steel from rusting. This is dependent on both the concrete strength (F'_c) and the cover

95

provided to the steel nearest to the surface. The Australian code requirements for minimum cover are given in Table 7.2

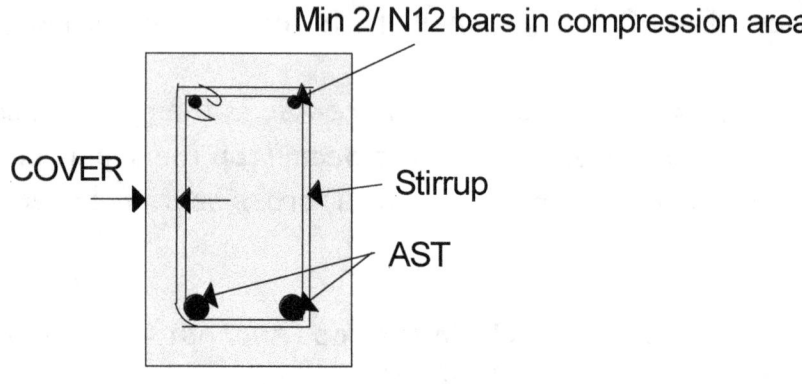

Table 7.2   Required Cover in mm

Exposure Class			F'c = 20 MPa	F'c = 25 MPa	F'c = 32 MPa
Interior	Typical	A1	20	20	20
Exterior	Inland >50km	A2	50	30	25
Exterior	Near Coastal <50 km	B1		60	40
Exterior	Coastal <1 km	B2			65

7.2.6 Anchorage of Steel and Lap Splices

Bars need to be anchored for a development length past the point at which they are no longer required. In the case of continuous beams this point for the top steel over supports is the "point of contraflexure" (point where moment changes from negative to positive). At least a third of the top steel should be continued past this point. At simply supported ends steel should be extended 12 bar diameters past the face of the support.

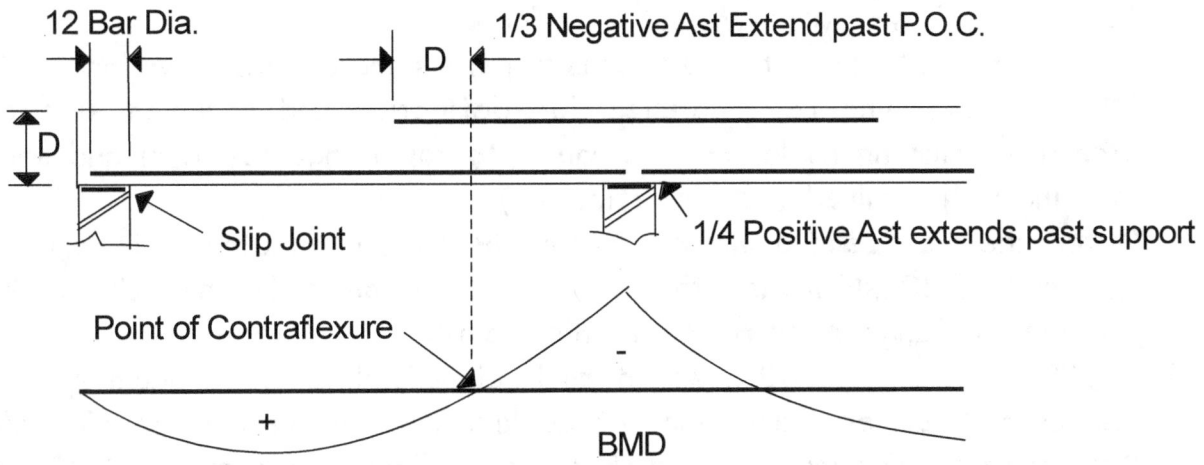

Where lapped bars are required to transfer the full tension between bars (eg where retaining wall starter bars are lapped with the main steel in the wall) lap lengths are of the order of 40 bar diameters.

Reinforcing mesh in slabs should be lapped such that the outer bar of one sheet of mesh laps at least two bars of the other sheet.

EXERCISES

1. A 250 UB37 spans 4.5 metres and carries a dead load of 20 kN/m and a live load of 10 kN/m. The top flange is laterally restrained in the centre. Check whether the moment capacity of this beam is sufficient and determine the deflection.

2. A 310 UB 46 spans 8 metres and has a lateral restraint to the top flange in the centre. What dead load can it carry for strength (in addition to its own dead load) and what is the deflection under the total dead load ? Use load factor of 1.35. Put $M^* = \phi M_b$ and solve for w.

3. A two span continuous steel UB spans 5 metres and carries a live load of 30 kN/m. Determine required size for deflection and required size for strength assuming no lateral restraints to top flange (by trial and error starting with required size for deflection).

4. Determine the size of a universal beam (UB) to span 6 metres if it carries a live load of 10 kN/m and the limiting deflection is 20 mm (Span/300). Assume top flange is restrained at third points.

5. A 600 mm wide by 500 deep internal reinforced concrete beam spans 6 metres and carries a dead load of 20 kN/m and a live load of 20 kN/m. Determine the required Ast and check for deflection, assuming 4/N20 bars in the top. Assume N24 bars, R10 stirrups and F'c = 25 MPa.

6. In question 5 if there were 5/N16 bars in the top and the depth was 450 mm instead of 500 what would be the required precamber ? Precamber is when you curve the formwork upwards so that the long term deflection = calculated deflection – precamber.

7. A 270 wide reinforced concrete lintel beam (exposed to the weather) spans 4 metres and carries a dead load of 12 kN/m (in addition to its own D.L.) and a live load of 15 kN/m. Choose a suitable overall depth (round up to nearest 50 mm) and calculate A_{st} required. Check deflection assuming 2/N16 bars in the top. Assume N20 bars in bottom, R10 stirrups and near coastal (<50km), F'c = 32 MPa.

8. PRESTRESSED CONCRETE, BRICKWORK AND GLASS

8.1 Prestressed Concrete

Prestressed concrete slabs and beams are used in preference to reinforced concrete beams when spans are long and dead loads are the predominant load.

Consider the case of a simply supported reinforced concrete beam subject to a live load in addition to its dead load. The combination of these produce a moment of $(w_{DL}+w_{LL}) \times L^2/8$ which is resisted by the concrete in combination with the reinforcement. Because concrete is weak in tension the concrete around the reinforcement cracks.

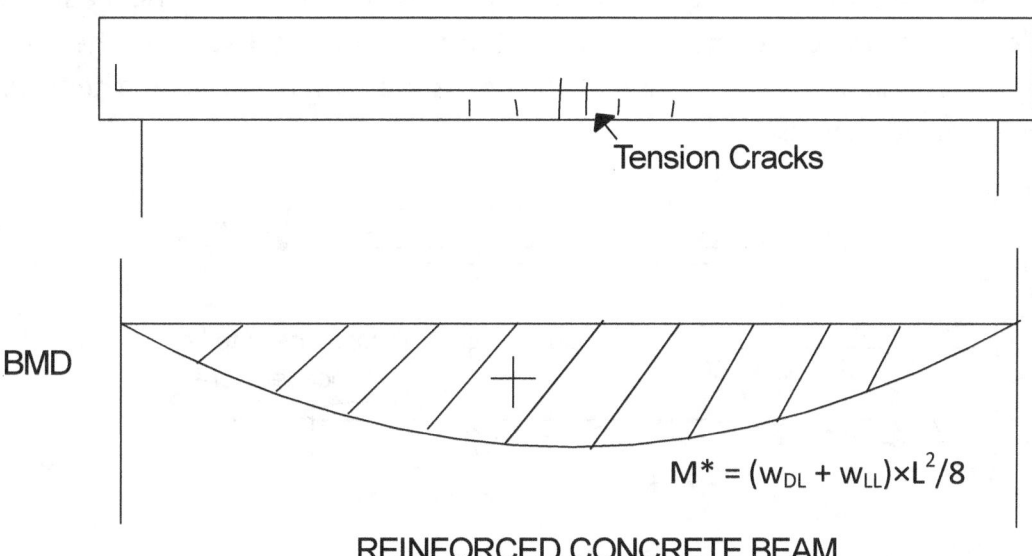

Tension Cracks

BMD

$M^* = (w_{DL} + w_{LL}) \times L^2/8$

REINFORCED CONCRETE BEAM

If we have draped prestressing cables instead of plain reinforcing steel with maximum eccentricity in the centre of "e" and prestressing force "P", this imparts a negative moment to the beam which is zero at the ends and a maximum (=P×e) at the middle. The resultant moment (addition of positive moment due to load and negative moment due to prestressing) is parabolic with a slight maximum positive moment in the centre. Because the concrete has a compressive prestress this negates the tension stress caused by the residual bending moment and thus there is no cracking.

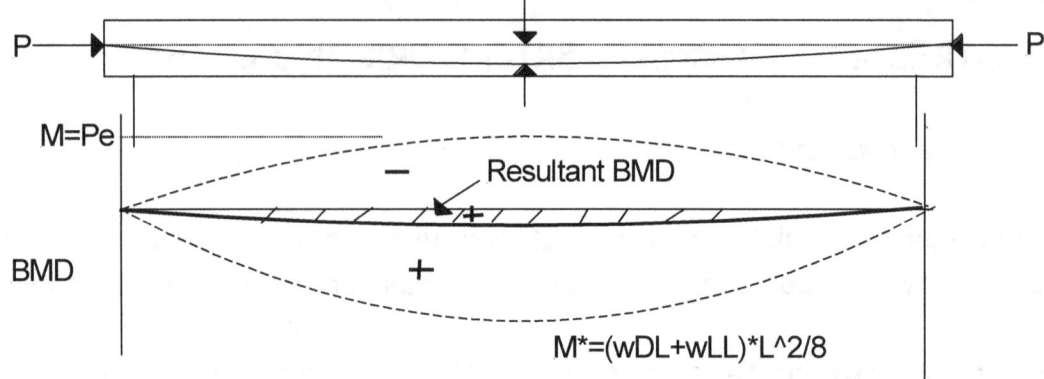

PRESTRESSED CONCRETE WITH DRAPED TENDONS

If we have straight prestressing tendons instead they provide an eccentric compressive force to the ends of the beam which is equivalent to applying a constant "negative" moment (=P×e) along the length of the beam. The resultant BMD has little or no value in the centre and high values (=P×e) at the ends. Normally however tendons are "debonded" at the ends which reduces these end moments to an acceptable value.

of

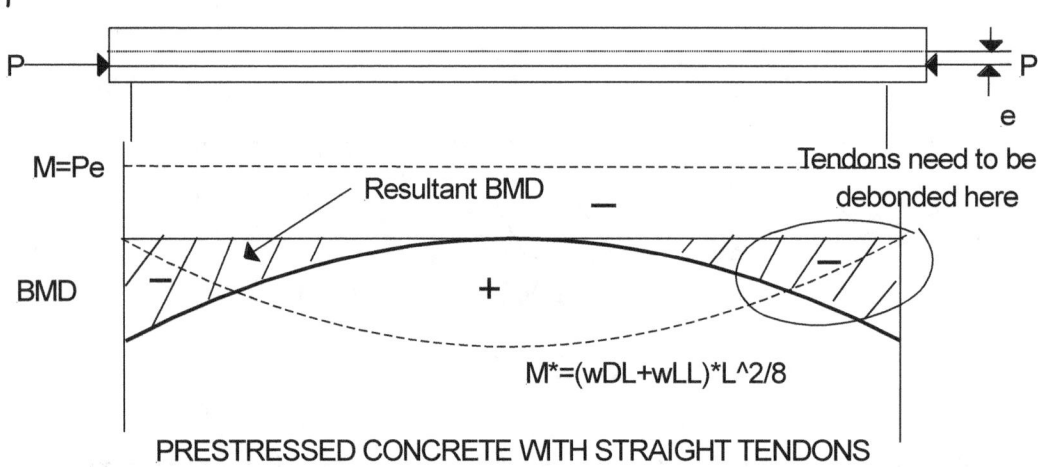

PRESTRESSED CONCRETE WITH STRAIGHT TENDONS

Typically prestressed concrete beams or slabs are stressed in two stages, one when the concrete strength is around 7 MPa and one when the concrete strength is around 75% of F'c.

8.2 Brickwork

Structural considerations for brickwork are

1. The compressive strength of the wall must be sufficient to carry the load. Commonly bricks have a compressive stress of 20 MPa but when combined with mortar in a wall the ultimate strength of the brickwork is 6.7 MPa for a 1:1:6 mortar. The other factor to consider is the height to thickness ratio which should not be greater than 6 for free standing walls nor 27 for walls supported laterally at the top. At a height of 8 times the thickness the compressive strength of a wall is 0.67 times the ultimate strength and at 27 it is about 25%.

2. Walls must be capable of transmitting wind loads to cross walls. This means that we have to assume some tensile capacity in the mortar (note that this is the only case where we do – we don't for brick retaining walls). Walls are checked for wind in the same manner that reinforced concrete floor slabs are. A bending moment is calculated and this is compared to the bending moment capacity of the brickwork assuming a failure tensile stress equal to the modulus of rupture" of the brickwork of 0.2 MPa.

3. Sufficient expansion joints must be provided to cater for the long term expansion that occurs once bricks leave the kiln. In NSW the 5 year expansion is around 1.1mm/m for extruded bricks and in Queensland it is about 0.6 mm/m. About half of this expansion will occur in the first year and the additional amount that occurs after 5 years is normally ignored, although it can be an additional 50%. Brickwork joint spacing is based on these amounts and an additional 0.35mm/m for thermal movement. So for the average brick in Queensland the expansion at each joint would be 0.6/2 = 0.3 mm/m and the total movement would be (0.3+0.35) =0.65mm/m. If joints are 15mm thick then the required spacing would be 15/0.65 = 23 m.

8.3   Glass

Ordinary glass is called annealed glass. The practical tensile strength of annealed glass is around 50MPa.   Note that annealed glass suffers a static fatigue effect which makes it only half as strong under long term loads (aquarium loads, snow loads, etc.) as under short term loads, such as wind gusts. A ϕ factor of 0.67 is used for strength checks.

For structural purposes such as where human impact is possible (windows next to doors or balustrades) safety glass is used. Safety glass is either
   a) Laminated glass – where two or more layers of annealed glass are bonded together. Laminated glass remains intact when broken due to the bonded layer. Laminated glass is 50% to 100% as strong (depending on aspect ratio and framing details) as monolithic glass of the same overall thickness and size when subjected to short duration loads at room temperatures.
   b) Toughened glass.- heat treated glass which has high strength and which breaks into small pieces on impact (eg car windscreen). Toughened glass is made from annealed glass via a thermal tempering process. The glass is placed onto a roller table, taking it through a furnace that heats it to above its annealing point of about 600 °C. The glass is then rapidly cooled with forced draughts of air while the inner portion of the glass remains free to flow for a short time. The strength of toughened glass is about 4 times that of ordinary annealed glass. Note that toughened glass can break without warning due to the presence of nickel sulphide inclusions. Heat soaking of toughened glass reduces the risk but it should not be used on awning roofs where the distance to the ground is above 3 metres nor for balustrades where the fall is greater than 1 m
   c) Laminated toughened glass –used for maximum safety.

The strength of annealed glass panels under wind pressure p_u can be checked in a similar manner to that of reinforced concrete slabs. First calculate M* and then compare it with ϕM_u where

$M^* = K \times p_u \times L_x^2$ where
L_x = short span (m) and
$K = 0.031 \times L_y/L_x$ with a maximum of 0.125 (equivalent to $wl^2/8$ for one way span when $L_y/L_x > 4$)
$\phi M_u = \phi \times fy \times Z/10^6$
$\phi = 0.67$
$fy \approx 55$ MPa
$Z = 1000 \times t^2/6$ mm^3
t = thickness of glass (mm)

For example consider a 1.94 metre by 1.55 metre glass panel which is 6 mm thick and subjected to an ultimate wind pressure of 2.3 kPa.

Then $L_y/L_x = 1.25$ and $K = 0.0388$
$M^* = 0.0388 \times 2.3 \times 1.55^2 = 0.21$ kN.m/m

$\phi M_u = \phi \times fy \times Z/10^6 = 0.67 \times 55 \times (1000 \times 6^2/6)/10^6 = 0.22$ kN.m/m

Since $\phi M > M^*$ the window is OK

(Referring to AS1288 "Glass in Buildings" we get an allowable wind load of 2.4 KPa for such a pane of glass)

The modulus of elasticity of glass is 70 GPa. (70,000 MPa). This can be used to calculate the deflection. Allowable deflection is normally $L_x/60$

Exercise:

1. According to AS1288 a 750 by 1312 annealed glass window needs to be 4 mm thick for a wind load of 3 KPa. Check it out using the simplified approach given above.

9.   MULTI-STOREY FLOORING SYSTEMS

9.1   Introduction

Flooring systems in multi-storey buildings consists of concrete floor slabs supported by beams which are in turn supported by columns, the latter transferring the dead and live loads of the floor to the footings.

Floor slabs can be
  a) Reinforced concrete or prestressed slabs spanning one way between beams. The slabs may be made up of a series of "T" beams sections joined together (rather than having a constant thickness) in order to reduce dead weight.
  b) Reinforced concrete or prestressed concrete spanning in two directions (two ways) between supporting beams. Reinforced concrete "waffle" slabs are also used sometimes to reduce dead weight, but are more expensive to build.
  c) Precast prestressed concrete planks, typically with tubular holes running in the middle of the slab in the direction of the span. These are usually "topped" with an insitu concrete slab which may in some cases give the floor extra strength by composite action.

Beams supporting floor slabs may be
  a) Steel beams or shallow steel trusses. Steel beams commonly have steel studs welded to the top to engage the concrete and produce a "composite beam" which is much stiffer and stronger than the steel beam alone.
  b) Reinforced or prestressed concrete beams which may be single span or continuous (spanning over three or more columns). Typically in car park or low rise commercial buildings shallow reinforced or more commonly prestressed (band) beams span between columns.
  c) Contained within the thickness of the slab. These flooring systems are called "flat plate" floors when there are no thickenings (drop panels) around the supporting columns and "flat slab" floors when there are. The floor in this case spans two ways and is divided into "column strips" which are effectively the beams and "middle strips". Column capitals are sometimes used in flat slab floors to prevent the slabs punching through the columns (punching shear).

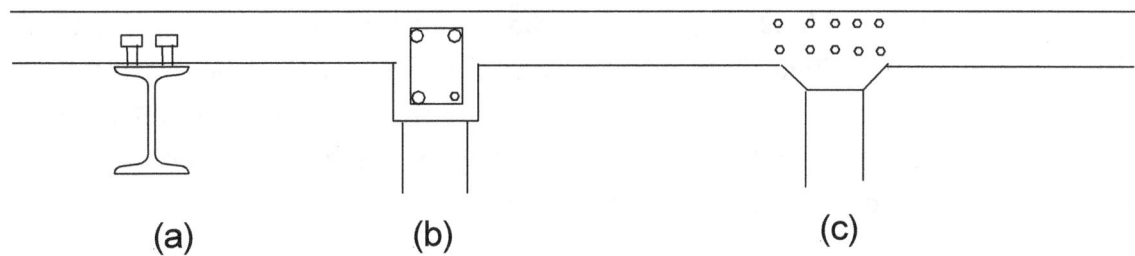

(a)   (b)   (c)

Typical overall span (L) to Overall Depth (D) ratios for initial sizing of slabs and supporting beams can be taken from Table 9.1 below but actual required thickness will be dependent on the live loading to be carried.

Table 9.1   Span to Depth Ratios for Sizing Elements

Element	L/D
Reinforced Concrete Slabs	
Simple Supported One Way	22
Continuous One Way	28
Simply Supported Two Way(L_y/L_x >1.4)	26
Simply Supported Two Way(L_y/L_x <1.4)	30
Continuous Two Way (L_y/L_x >1.4)	30
Continuous Two Way (L_y/L_x <1.4)	34
Cantilever	9
Prestressed Concrete Slabs	
Continuous One Way with Band Beams	40
Flat Plate	40
Precast Concrete Planks	40
Prestressed Concrete Beams	
Continuous Band Beams	25

9.2   Design of Simply Supported One –Way R.C. Slabs

One way slabs are designed as one metre wide strips spanning between supports. They are designed in a similar manner to reinforced concrete beams. Bending moments are calculated for each metre wide strip and the required area of steel (A_{st}) is then calculated.

From Section 7.2.3

$$Ast = 2650 \times M^*/d$$

Where M* = Design moment per 1m width of slab and
d =Depth from top of compression face to centre of steel

Steel Designations   Mesh SL - equal diameter bars at equal spacing in both directions (eg SL 72 is 7mm diameter bars at 200mm centres both ways) or RL – thicker bars one way at typically 100 centres with 8mm cross bars at 200 mm centres (eg RL 1218 has 12 mm bars at 100 mm centres in one direction and 8 mm bars at 200 mm centres in the other direction)

Bars as for beams. Bars with a yield stress of 500 MPa are given the prefix "N" , eg N12 is 12 mm dia bar with a yield stress of 500 MPa. Common bar sizes are N12 (A_b=113mm^2), N16 A_b=201mm^2, N20 (A_b=314mm^2

Limit on p (A_{ST}/bd) of 0.005 (0.5%). Note that for slabs we consider a nominal metre wide strip and therefore b = 1000 mm

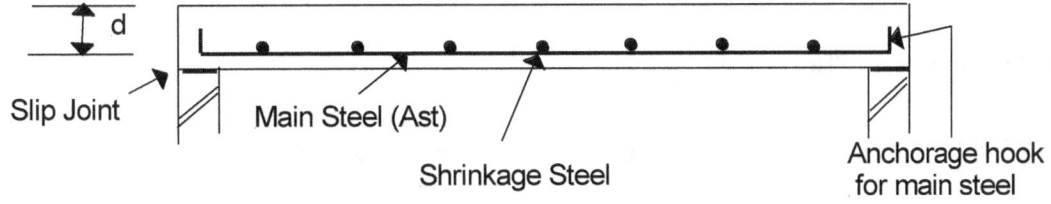

The effective depth of the slab (d) is equal to the overall depth (D) minus the required concrete cover minus half the bar diameter.

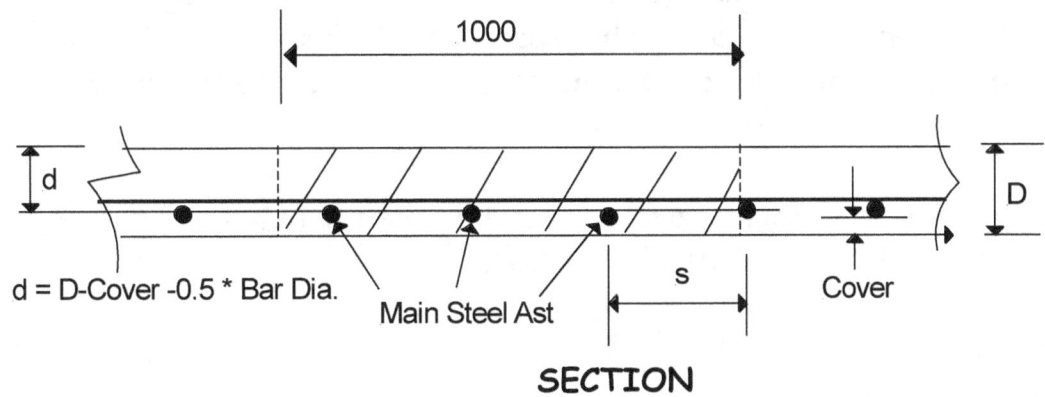

SECTION

Required bar spacing(s) =Cross sectional Area of one bar×1000/ A_{st}(mm^2/m)

Deflection is checked based on a cracked moment of inertia (I) of 0.033bd^3 with a load equal to dead load plus one third of the live load.   Allowable deflection is Span/300. Modulus of elasticity (E) is 5055×F'c$^{0.5}$.

For long term deflection we may assume a creep factor of [3-1.2× A_{sc}/A_{st}] where Asc is the cross sectional area of steel in the compressive zone of the beam (top for simply supported members) and Ast is the cross sectional area of steel in the tension zone.

Example: Consider the case of a one way slab required to span 5 metres and carry a live load of 4.9 kPa. Required cover is 20 mm. We need to find the required slab thickness and the required spacing of reinforcing steel (assume 12 mm bars) and check long term deflection. F'c = 40 MPa.

Assume as a first guess a slab thickness of 230 mm (~Span/22 = 5000/22 = 227).

Then Dead load = 0.23× 25 = 5.75 kPa
Total Load per metre wide strip of slab (w^*) =1.2×5.75+ 1.5×4.9 = 14.25 kN/m

Max Bending Moment (M*) = 0.125×14.25×5^2 ($wL^2/8$)= 44.5 kN.m/m
Effective depth of slab is 230 – 20 -6 = 204 mm
Using the simplified formula required area of reinforcing steel is
Required A_{st} = 2650×44.5/204 = 578 mm^2/m

Check that Ast/bd is less than 0.005

In this case 578/(1000×204) = 0.003 <0.005 so depth is OK

Assuming N12 bars area of one bar ~ π×6^2 = 113 mm^2
Required spacing of N12 mm bars is then 113×1000/578 = 196 mm say N12@190 mm

You will also need steel at right angles to the main steel for shrinkage, typically about 0.17 % of the gross area, or in this case 0.0017×1000×230 = 391 mm^2/m. Required spacing of N12 bars (Area ~ 113 mm^2) would be 113×1000/391 = 289 mm say N12@250 mm

Now for deflection check . Assume say SL72 mesh in the top (A_{sc} = π×3.5^2/0.2 = 192 mm^2/m) and actual A_{st} = 113/0.19 = 595 mm^2/m.

Then A_{sc}/A_{st} = 192/595 ≈ 33%)
Creep Factor = 3 -1.2×0.33 = 2.6

Load for deflection check = 5.75 + 0.33×4.9 = 7.37 kN/m
I = 0.033×b×d^3 = 0.033×1000×204^3 = 280×10^6 mm^4
E = 5055 × $\sqrt{}$ F'c = 5055 × $\sqrt{40}$ = 31,970 MPa
Deflection ~ 5×w×L^4/(384×E×I) × Creep Factor
            = [5×7.37×5000^4/(384×31,970×280×10^6)] × 2.6
            = 17 mm

Assuming an allowable long term deflection of Span/300 gives an allowable deflection of 17 mm and therefore the slab is OK for deflection.

9.3   Design of Continuous One -Way R.C. Slabs

Continuous one way reinforced slabs are designed in exactly the same manner as simply supported slabs except that the bending moments and deflection are less than would be the case for a similar simply supported span. For two span beams with uniformly distributed loading (w) the design moments are different to those shown in Section 4.9 ($0.125wL^2$ for -ve moment and $0.071wL^2$ for positive moment) due to the ability of reinforced concrete to flex and reduce moments thus decreasing the negative moment and increasing the positive moment (this is called moment redistribution).

For design of continuous beams and slabs with more than two spans the Maximum Bending Moment is taken as $0.10 \times wL^2$ for negative (top) steel over supports and $0.091 \times wL^2$ for positive (bottom) steel in the end span and $0.063 \times wL^2$ for positive (bottom) steel in interior spans. Effective depth (d) is measured from the top face to the centre of the top steel for steel over supports (negative steel)

Deflection is calculated for worst case (end span) as $1/185 \times wL^4/EI$ which is the same as that given in Section 4.9 for two span beams.

9.4   Design of Two Way Slabs

Two way slabs are designed in much the same way as one way slabs except that it is assumed that one of the one metre wide strips carry a certain proportion of the dead and live load in one direction and the corresponding one metre wide strip at right angles carries the remainder. The shorter span always carries the greatest load and for that reason the steel closest to the bottom should run in the short direction (for tension on the bottom or positive moment). Similarly for negative moments the steel closest to the top should run in the short direction.

For simply supported two way slabs the proportion carried in each direction can be calculated on the basis that each strip must deflect the same amount as shown below.

Moments to be designed for in each direction are equal to a bending moment coefficient (See Table 9.2) multiplied by the factored total dead and live load pressure (w*) by the span squared. For example in the case of a slab which is simply supported on four sides with the long side (L_y) 1.5 times the short side (L_x) the short span moment is $0.093w*L_x^2$ whilst the long span coefficient is $0.025w*L_y^2$. For very skinny slabs ($L_y \ggg L_x$) the moment in the short direction would be $0.125w*L_x^2$ ($w*L_x^2/8$) which is the value for a simply supported slab spanning one way.

The other major difference with two way slabs is in the calculation of the effective depth (d). Note that d is different for each direction due to the different layers of steel. Normally the bottom steel is placed spanning the short direction so for the short direction

d_x = D – Cover – ½×Bar dia.

and for the other direction (normally long direction)

d_y = D – Cover –1½×Bar dia.

For slabs with continuous edges there will be top (negative) moments over the supports similar to that required for continuous beams. We will assume that the top cover in this case is the same as the bottom cover and therefore "d" for the top steel will be the same as the "d" for the bottom steel in each direction.

Table 9.2   Bending Moment Coefficients for Two Way Slabs

Ly/Lx	1.0	1.1	1.2	1.3	1.4	1.5	1.75	2.0
4 Edges Simply Supported								
Short Span Bottom	0.056	0.066	0.074	0.081	0.087	0.093	0.103	0.111
Long Span Bottom	0.056	0.046	0.039	0.033	0.029	0.025	0.018	0.014
4 Edges Continuous								
Short Span Bottom	0.024	0.028	0.032	0.035	0.037	0.04	0.044	0.048
Long Span Bottom	0.024	0.020	0.017	0.014	0.012	0.011	0.008	0.006
Short Span Top	0.032	0.037	0.043	0.047	0.049	0.053	0.058	0.064
Long Span Top	0.032	0.027	0.023	0.019	0.016	0.015	0.011	0.008

Example:

A 7 metre by 4 metre slab is simply supported on all sides, i.e. four edges discontinuous. It carries a live load of 4.5 kPa, F'c = 32 MPa, cover =30 mm (exterior inland) and N12 bars are to be used as reinforcement.

L_y/L_x = 7/4 = 1.75. From Table 9.1 first guess for thickness = Short Span/26 = 4000/26 = 154 say 150 mm. This will have to be checked to see that the calculated steel percentage is less than 0.5% and also that the deflection is OK. If not a thicker slab may be required.

The required spacing of the N12 bars in each direction is calculated as follows.

w^* = 1.2×0.15×25 + 1.5×4.5 = 11.25 kPa.

Short Direction:

d_x = 150-30-6 = 114 mm

M_x^* = 0.103×11.25×4^2 = 18.54 kN.m/m

A_{st} = 2650×18.54/114 = 431 mm^2/m

Steel percentage = 431/(1000×114) = 0.004

Area of one N12 bar = 113 mm^2

Required bar spacing in x direction = 1000×113/431 = 262 mm say N12@260

Long Direction:
d_y = 150-30-12-6 = 102 mm
M_y^* = 0.018×11.25×7^2 = 9.92 kN.m/m
A_{st} = 2650×9.92/102 = 258 mm^2/m
Area of one N12 bar = 113 mm^2
Required bar spacing in x direction = 1000×113/258 = 439 mm say N12@430
However it is normal to have N12 bars at a maximum spacing of 400mm, so choose N12@400.

Using the "Slabs" program (See Section 9.5) the deflection of this slab is 13.1 mm which is less than the allowable of 4000/300 = 13.3 mm

Reinforcing plan would look something like this. B1 denotes these bars are placed first with B2 bars on top. Note that short span bars should always be closest to the bottom.

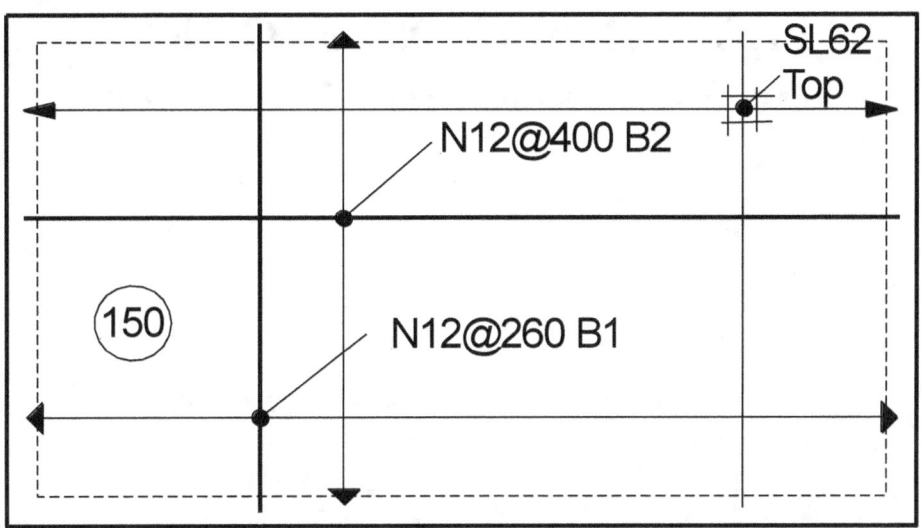

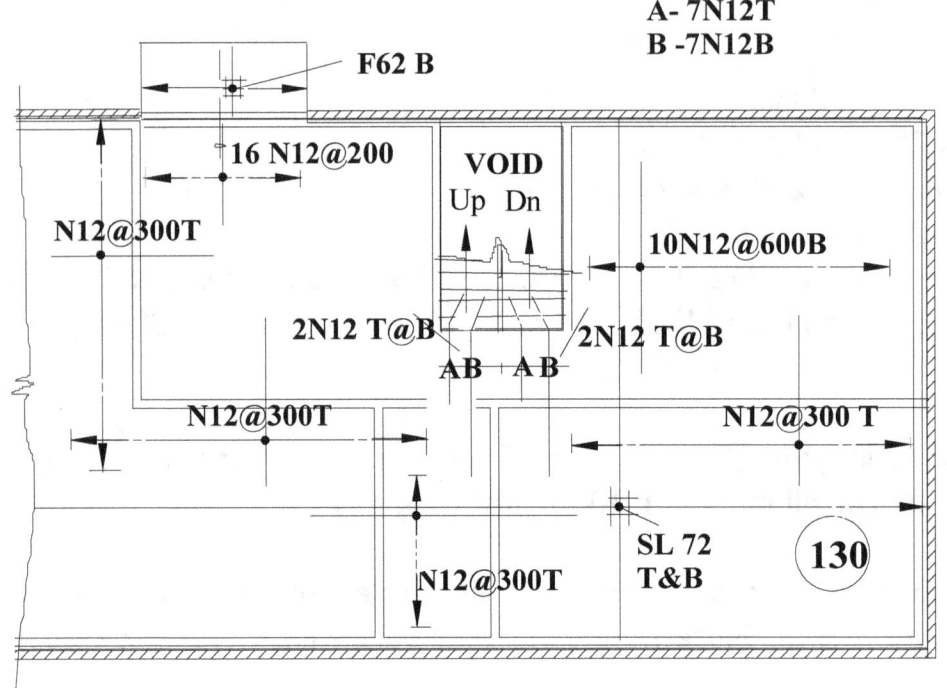

A- 7N12T
B -7N12B

F62 B

16 N12@200

VOID
Up Dn

N12@300T

10N12@600B

2N12 T@B   2N12 T@B
AB   AB

N12@300T

N12@300 T

SL 72
T&B

130

N12@300T

PLAN OF SECOND FLOOR

Figure 9.1 Layout of Steel in Two-Way Slab

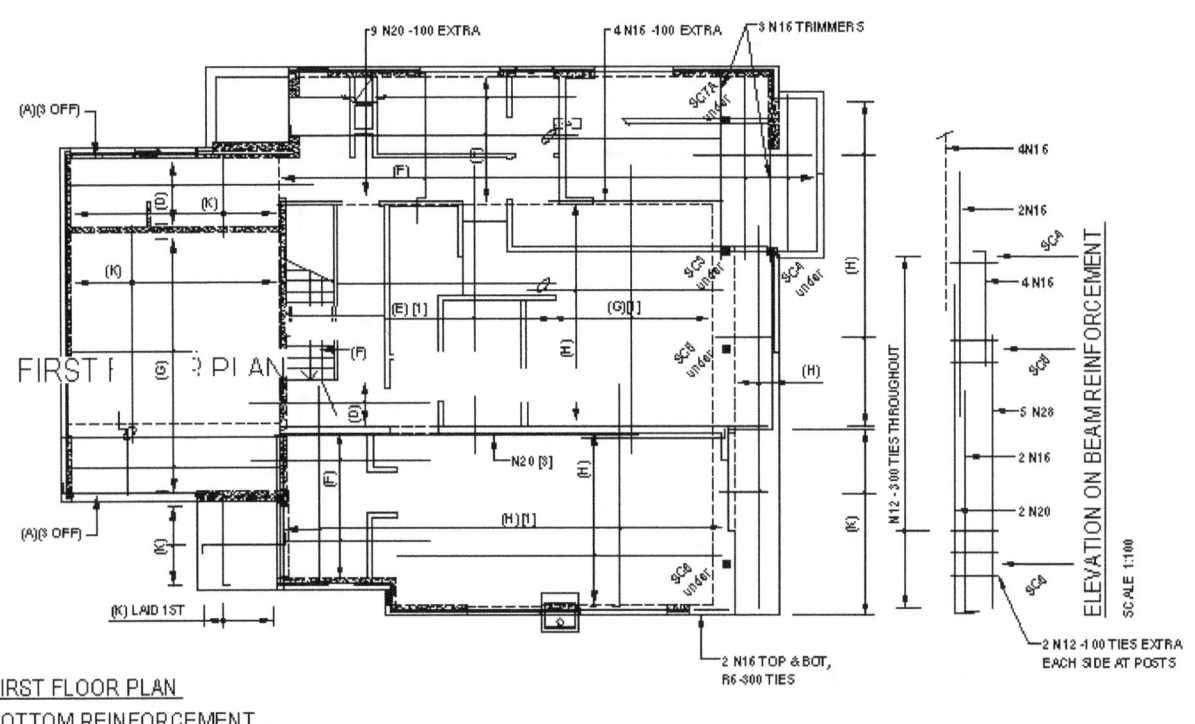

9 N20 -100 EXTRA   4 N16 -100 EXTRA   3 N16 TRIMMERS

(A)(3 OFF)

(F)

(D)   (K)

(K)

(E) [1]   (G)[1]

FIRST F   R PLAN

(G)   (F)

(A)(3 OFF)

(K)

(K) LAID 1ST

N20 [3]

(H)

(H)[1]

SC7A under

SC3 under

SC4 under

SC8 under

N12 -300 TIES THROUGHOUT

(H)

(H)

(K)

4N16

2N16

SC4

4 N16

SC8

5 N28

2 N16

2 N20

SC4

ELEVATION ON BEAM REINFORCEMENT

SCALE 1:100

2 N16 TOP & BOT,
R6 -300 TIES

2 N12 -100 TIES EXTRA
EACH SIDE AT POSTS

FIRST FLOOR PLAN
BOTTOM REINFORCEMENT

SCALE 1:100

BOTTOM REINFORCEMENT

SCALE 1:100

SLAB REINFORCEMENT SCHEDULE

BAR GROUP TYPE	REINFT	BAR GROUP TYPE	REINFT	BAR GROUP TYPE	REINFT
(A)	N12-50	(A16)	N16-50	(A20)	N20-50
(B)	N12-100	(B16)	N16-100	(B20)	N20-100
(C)	N12-120	(C16)	N16-120	(C20)	N20-120
(D)	N12-150	(D16)	N16-150	(D20)	N20-150
(E)	N12-180	(E16)	N16-180	(E20)	N20-180
(F)	N12-200	(F16)	N16-200	(F20)	N20-200
(G)	N12-220	(G16)	N16-220	(G20)	N20-220
(H)	N12-250	(H16)	N16-250	(H20)	N20-250
(J)	N12-280	(J16)	N16-280	(J20)	N20-280
(K)	N12-300	(K16)	N16-300	(K20)	N20-300

PROVIDE N12-300 CROSS RODS TO TIE UN-SUPPORTED
MAIN REINFORCEMENT BARS SHOWN ON PLAN

——— BARS LAID FIRST [UNO]

Figure 9.2   Layout of Steel in Two-Way Slab (Another Example)

9.5   Design of Slabs using "Slabs" program

The author was a co-developer of the Slabs concrete program which is marketed by Inducta Engineering (www.inducta.com.au). This program is based on a finite element analysis of slabs with a graphical input of geometry, loads and material properties. The program calculates bending moments and deflections and can accommodate any system of supports (columns, walls etc)

The screen shots below show the bending moment contours for a 150 mm thick 8 m square panel, simply supported on all sides and with a live load of 2 kPa.

From above

Factored Dead Load is 1.2×0.15×25 = 4.5 kPa
Factored Live Load is 1.5×2 = 3 kPa

Therefore w* = 7.5 kPa and from Table 9.2
$M_x^* = 0.056 \times 7.5 \times 8^2 = 26.9$ kN.m
This can be compared to the Slabs output value of 25.6 kN.m

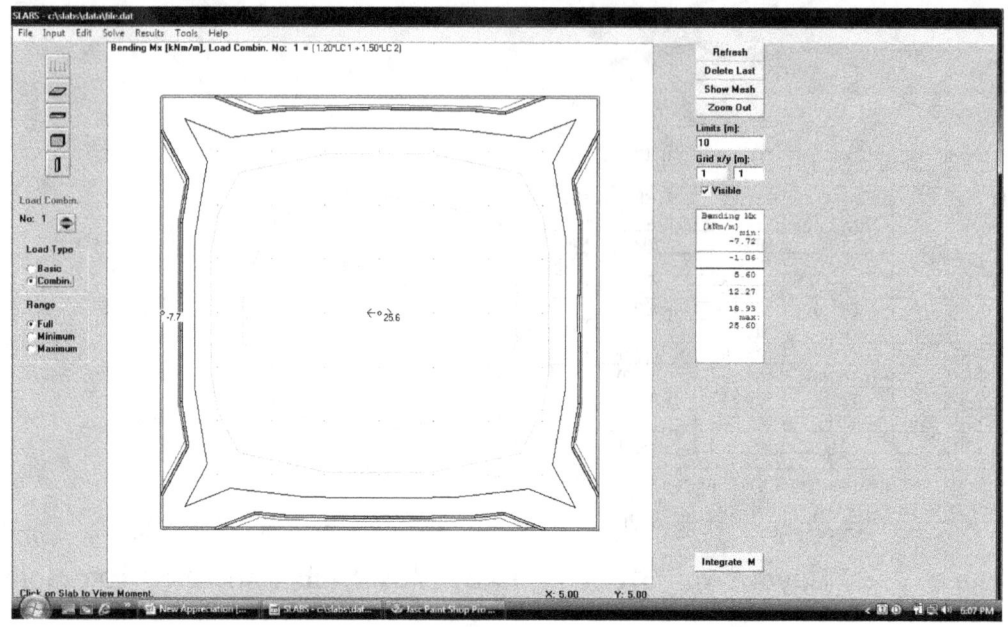

Figure 9.3 "Slabs" Program Output

9.6   Design of R.C. Flat Slab Floors

These are designed much the same way as two way slabs except that the slab is divided into "column" strips which are essentially beams (contained within the slabs) spanning between the columns and middle strips which are essentially two way slabs. The column strips are defined as one quarter of the span width either side the line joining the columns. Calculation of moments and deflections involve coefficients similar to those for two way slabs but will not be considered here.

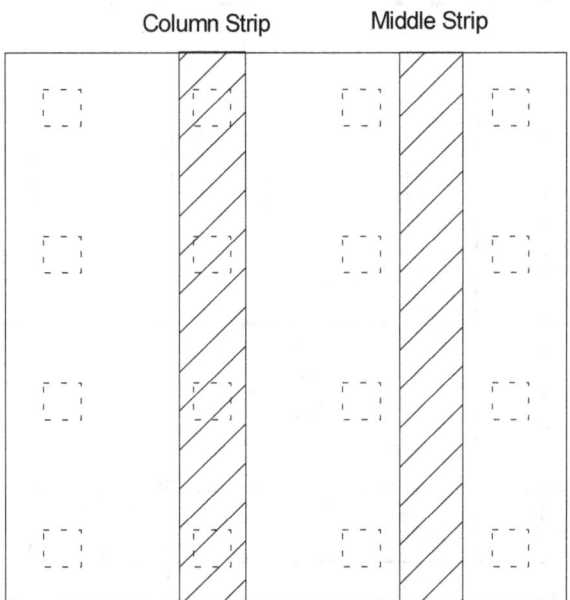

Prestressed concrete flat slab floors are similarly designed as column and middle strips and in some cases very few prestressing cables are required in the middle strips. Figure 9.4 shows the layout of prestressing cables in a flat slab floor. In this case there are two cables running north-south down the central column strio with a maximum drape of 125 mm (measured from the bottom of the slab) over columns.

115

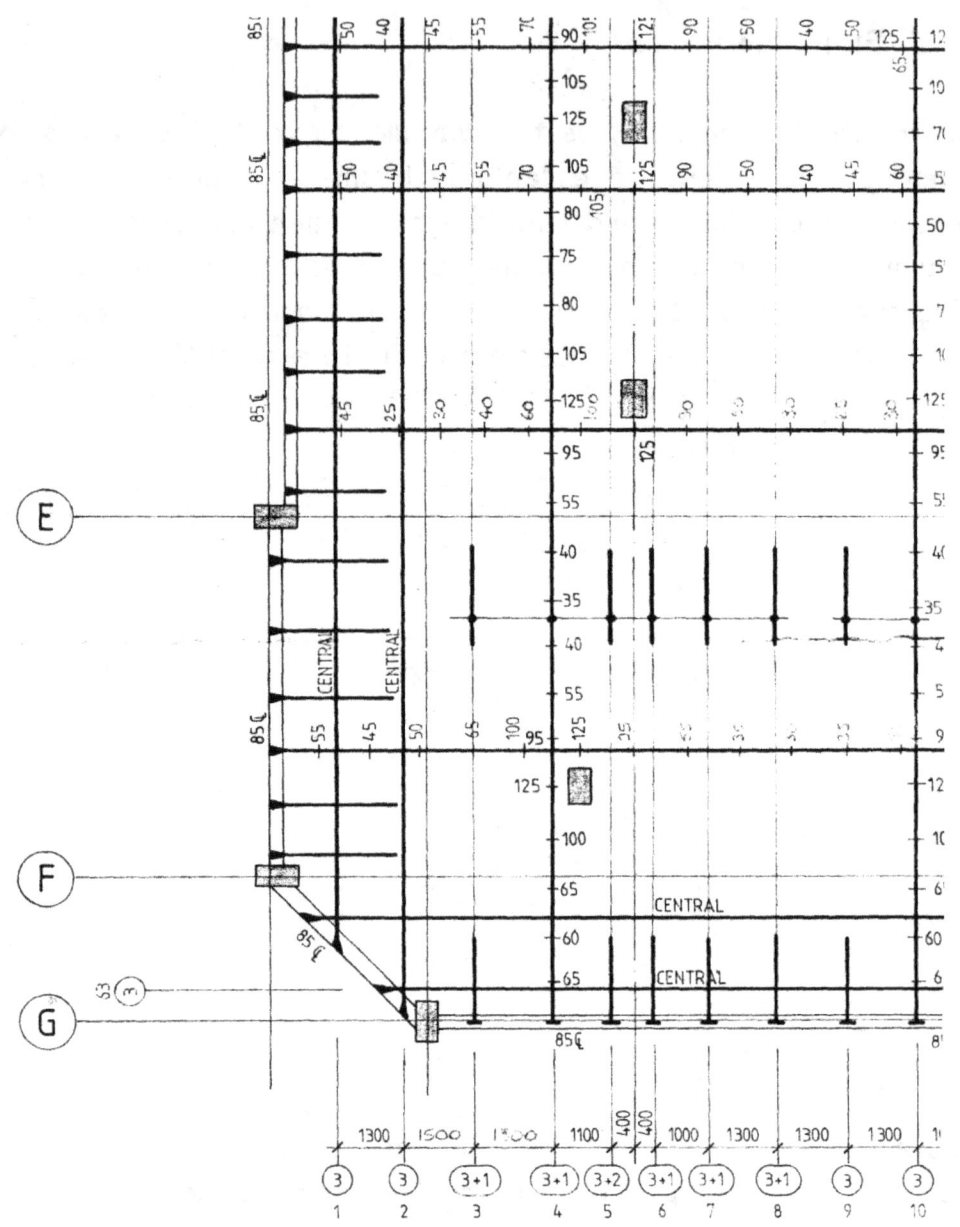

SECOND FLOOR PLAN - 170 THIC
TYPICAL 3RD - 7TH INC.

Figure 9.4   Prestressed Concrete Two Way Slab

116

9.7   Composite One Way Slabs and Beams

Composite steel beams utilize the help of the slab in both strength and deflection if the slab is adequately connected to the steel beam with shear connectors. Common types of composite floors are

9.7.1   Composite Slabs Using Permanent Steel Formwork

"Bondeck" and other similar profiles are steel sheeting which acts compositely with the concrete. Shear transfer between concrete and steel is achieved by means of ribs which stick into the concrete.   The composite concrete slabs span one way between supports.

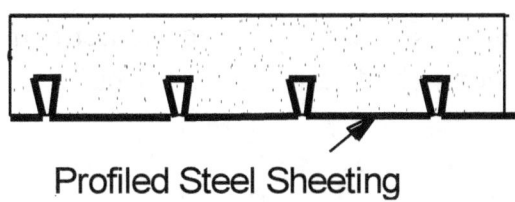

Profiled Steel Sheeting

Figure 9.5   Permanent Formwork Profile

These slabs do not require formwork as the steel sheeting supports the concrete load, although in some cases props are needed to cut down the length of unsupported spans.   Although the sheeting is galvanized corrosion of the unprotected steel, especially in coastal regions, can be a problem.

9.7.2   Composite Beams
Steel beams are often connected to the slabs they support by shear connectors in order to make the beams and slabs work compositely. The shear connectors are typically "studs" welded to the top of the steel beam at regular intervals.

Shear Connectors

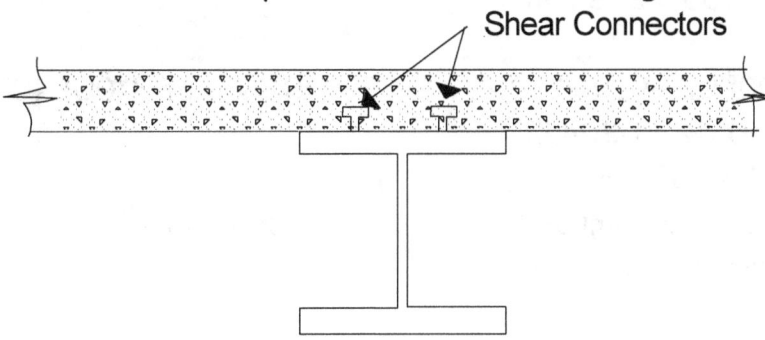

Figure 9.6   Composite Steel Beam and Slab

These studs must be strong enough to transfer the compressive forces that develop in the concrete to the steel beam. The strength of the composite section comes from the increased moment of Inertia and Section Modulus due to composite action.

9.7.3 Moment of Inertia of Composite Beams

The effective width of slab (b_{eff}) that acts compositely with the beams is equal to 16 times the depth of the slab plus the flange width of the beam. For example if a slab is 100 thick sitting on a 310 UB 40 (Flange width 165) then the effective width of slab is 16×100+165 = 1765 mm.
The effective with of the slab is then reduced in relation to the relative moduli of elasticity of the steel and concrete (N) to effectively change the concrete slab into a steel section. For 25 MPa concrete N is approximately equal to 15 which gives b_{eff} = 1765/15 = 118 mm in the example.

The position of the centroidal x-x axis and the moment of inertia of the section is then calculated in accordance with the method set out in Appendix C.

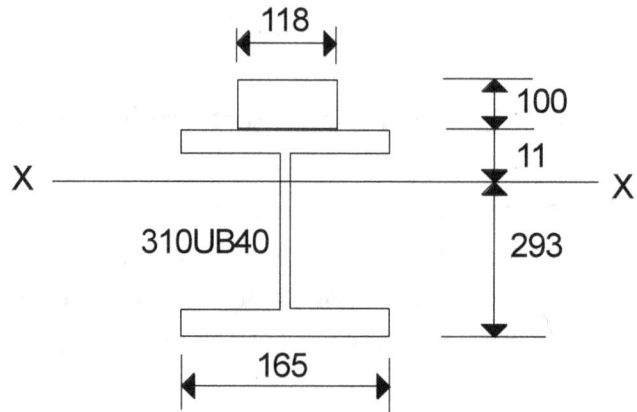

To calculate y_b
Total area = 118×100 +5150 (area of 310UB 40) = 16,950 mm^2
Moments about base = (118×100)×354 +5150×(304/2) = 4,960,000 mm^3
y_b = 4,960,000/16,950 = 293 mm

Then I_{xx} = 118×100^3/12 + (118×100)×61^2 + 85.2×10^6 (I_{xx} of 310UB40)
         + 5150 ×(293-304/2)^2
         = 241.3×10^6 mm^4
Note that the I of the composite section (241.3×10^6 mm^4) is 2.83 times the I of the steel beam alone (85.2×10^6 mm^4)

9.8 Post Tensioned Band Beams

Post tensioned band beam slabs are commonly used in multi storey buildings. They are essentially designed as one way continuous slabs spanning between continuous band beams. Cables are enclosed in cable ducts. Typically 4/12.5 mm cables are contained in an oval duct. Cable ducts are draped so that they are close to the top of the beams at the supports and close to the bottom of the beams at mid-span. The vertical distance between the centre of cables over the supports and the centre of cables midspan is termed the "drape" of the cable (e). For simply supported beams the drape is from the centroid of the beam to the centre of the cables in the bottom (See Figure 9.7)

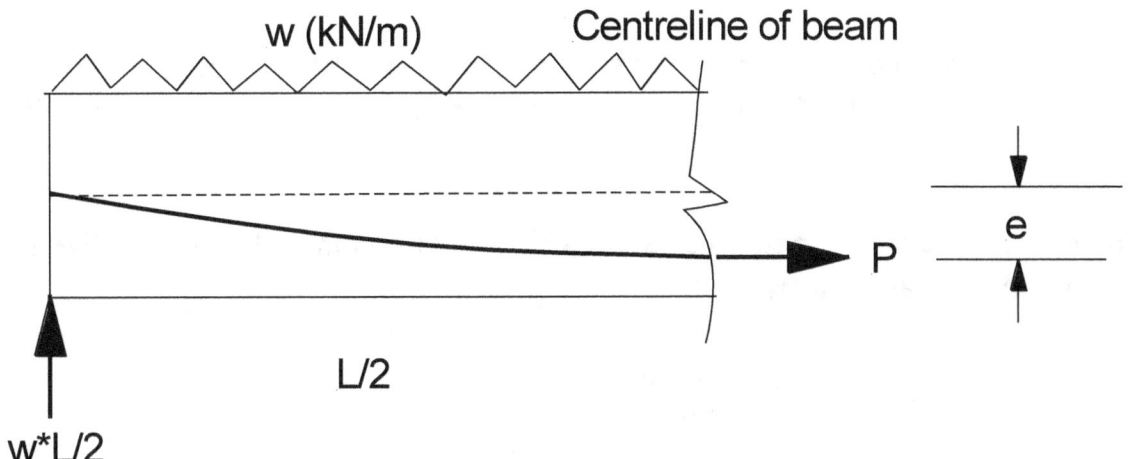

Figure 9.7 Free Body Diagram of Half of Beam

Taking half of the beam as a free body and taking moments about the centerline with w as the load balanced by the prestressing gives

wL/2×L/2 – w×L/2×L/4 –P×e = 0 or

Cable Force Required (P) = $wL^2/8e$

Consider the case of a band beam spanning 8.4 metres (Column grid 8.4 metres by 8.4 metres) with a depth of 350 mm. Further assume a drape of 250 mm (depends on cover)

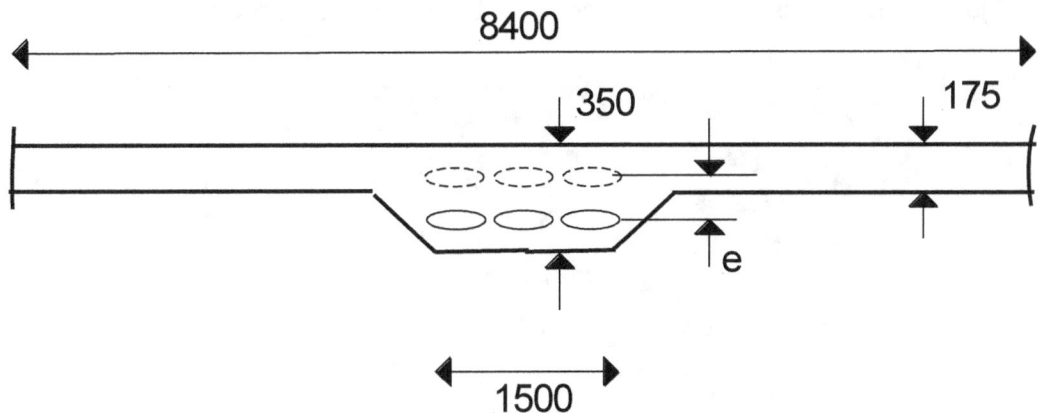

119

In this case the dead load supported by the band beam is 44 kN/m. Assuming a live load of 5 kPa we might choose to balance 10% of it or 0.5 kPa, giving a load to be balanced by the prestressing of (0.5×8.4)+44 = 48.2 kN/m to be balanced.

We then calculate the required prestressing force as

$$P = 48.2×8.4^2/(8×0.25) = 1700 \text{ kN}$$

Allowing for friction and stressing losses we may assume that a VSL 4 strand slab tendon can carry a load of 440 kN, which means that we will need 1700/440 = 3.86 say 4 tendons

We then need to check the beam for stress due to the remainder of the live load and calculate the losses exactly

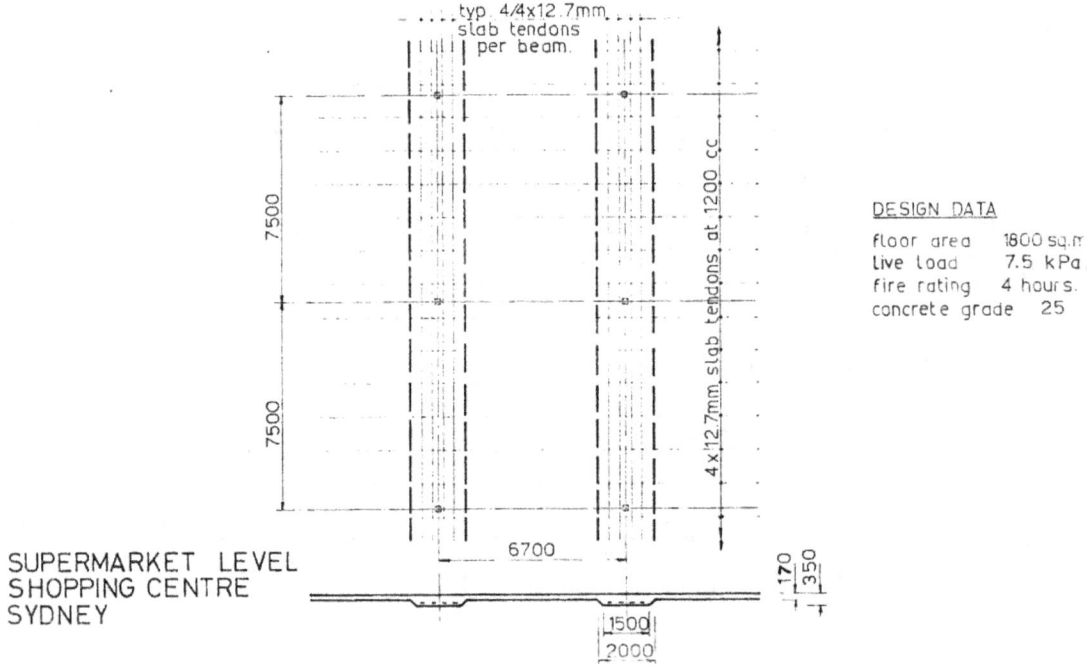

Figure 9.8 Prestressed Concrete Band Beam Layout

Figure 9.9   Fixing Anchorage Steel for Prestressing Cables
Source: Lysaght Bondek Manual

120

9.9 Formwork Stripping

9.9.1 Introduction

The question of when to strip formwork is frequently faced by Builders. In most cases they resort to consulting tables of stripping times published in various codes of practice but in some cases there may be a need to calculate stripping times if conditions are such that the published standard stripping times do not apply or are based on too conservative assumptions. For example in the Australian standard AS 3600 "Concrete Structures' stripping times are based on an estimated construction live load of 2 kPa. Where construction live loads are less than this considerable time can in some cases be saved by earlier stripping. This section sets out the basis for determining the required stripping time and relates this to the standard stripping times published in AS 3600.

9.9.2 Concrete Strength Gain with Time

The calculation of stripping times requires a knowledge of the variation of concrete strength with time and curing temperature. For normal strength concrete the factors shown in Table 9.1 should be multiplied by the 28 day compressive strength (F'_c) to get an estimate of the compressive strength of the concrete.

Table 9.1 Ratio of Compressive Strength (Fc) to F'c

Time (Days)	Temperature in Degrees Centigrade						
	5	10	15	20	25	30	35
1	0.07	0.10	0.13	0.16	0.19	0.22	0.24
2	0.16	0.22	0.27	0.32	0.37	0.41	0.44
3	0.24	0.32	0.39	0.44	0.50	0.54	0.58
4	0.32	0.41	0.48	0.54	0.59	0.64	0.68
5	0.39	0.48	0.56	0.62	0.67	0.71	0.74
6	0.44	0.54	0.62	0.68	0.72	0.76	0.79
7	0.50	0.59	0.67	0.72	0.77	0.80	0.83
8	0.54	0.64	0.71	0.76	0.80	0.83	0.86
9	0.58	0.68	0.74	0.79	0.83	0.86	0.88
10	0.62	0.71	0.77	0.82	0.85	0.88	0.90
11	0.65	0.74	0.80	0.84	0.87	0.90	0.92
12	0.68	0.76	0.82	0.86	0.89	0.92	0.93
13	0.70	0.78	0.84	0.88	0.91	0.93	0.95
14	0.72	0.80	0.85	0.89	0.92	0.94	0.96
18	0.79	0.86	0.90	0.93	0.96	0.97	0.98
21	0.83	0.89	0.93	0.96	0.97	0.99	1.00
24	0.86	0.92	0.95	0.97	0.99	1.00	1.01
28	0.89	0.94	0.97	0.99	1.00	1.01	1.02
60	0.99	1.02	1.03	1.04	1.04	1.05	1.05
365	1.06	1.06	1.07	1.07	1.07	1.07	1.07

For example if 30 MPa concrete is cured at a temperature of 10 degrees Centigrade it will have reached 71% (i.e. 21.3 MPa) of its 28 day strength (F'c) in 10 days .

9.9.3 Stripping Stages

Although it is possible to strip all of the formwork and supporting props in one go it is more normal when stripping before 28 days to first of all strip the soffit formwork between props and then remove the props at a later date. This is commonly referred to as two stage stripping.

The first stage is commonly done in the first week and the second stage after two or three weeks. The criteria for stripping for each stage is as follows.

First Stage Stripping

In this case the primary criteria is that the concrete between the props remain uncracked following removal of the soffit forms

This criteria can be expressed in Limit State format as

Design Ultimate Resisting Moment > Design Bending Moment

where

> Design Ultimate Resisting Moment = $\phi\,M_{uo}$ = $\phi \times f'_{cf} \times Z$
> $\phi\,M_{uo}$ = Cracking Moment Capacity
> ϕ = Strength Reduction Factor = 0.8
> f'_{cf} = failure flexural tensile strength
>   = $0.55 \times \sqrt{F'_c}$ (F'_c = Concrete Compressive Strength)
> Z = Element section Modulus

and

Design Bending Moment = M* = Applied moment due to factored loads using a load factor of 1.2 for dead loads and 1.5 for live loads.

Note that this criteria forms the basis for the stripping times given in AS 3600 (Table 19.6.2.4).

To explain how one would determine the required stripping time by calculation and to relate the times in Table 19.6.2.4 to the calculations, consider the case of a continuous slab spanning 2 metres between props. According to AS 3600 Table 19.6.2.4 would only apply if the concrete had a 28 day strength of at least 20 MPa, where the construction live load was 2 kPa and where the slab thickness was at least 100 mm, so these figures will be used in the following example.

f'_{cf} = $0.55 \times \sqrt{F_c}$ MPa

$Z = 1000 \times 100^2 / 6 = 1.67 \times 10^6$ mm^4

Therefore $\phi\,M_{uo}$ = $0.8 \times 0.55 \times \sqrt{F'_c} \times 1.67 \times 10^6 / 10^6$
            = $0.735 \times \sqrt{F'_c}$ kN.m

Now the factored load per unit length (w*) = $1.2 \times 0.1 \times 25 + 1.50 \times 2 = 6$ kN/m.

M* ≈ w* $\times$ L^2 / 10 (This is the maximum moment in continuous concrete elements with <u>more than 2 spans</u> and allowing for some moment redistribution)
     =$6 \times 2^2 / 10 = 2.4$ kN.m

Equating M * with ϕM_{uo} gives

$$0.735 \times \sqrt{F'_c} = 2.4 \text{ or Required } F'_c = 11 \text{ MPa.}$$

Now $F_c / F'_c = 11/20 = 0.54$ and from Table 9.1 above we see that at 20 degrees C we can strip after 4 days. (Note that Table 19.6.2.4 of AS3600 says that if the curing temperature is 20 degrees soffits can be stripped after 4 days.)

Second Stage Stripping

Note that if there are no props within the span of an element , i.e. if formwork spans between final supports, then the second stage stripping criteria applies.

The main striking criteria in this case are

1. The strength of the element must be assured.   This criteria is satisfied if the ratio of the concrete strength at stripping to the 28 day strength is greater than the ratio of the factored construction loads to the factored design loads.

    I.e. if F_c / F'_c > (1.2 × D L + 1.5 × Construction L.L.) / (1.2 × D.L. + 1.5 × Design L.L.)

2. The total long term deflection should not be significantly different to that assuming stripping at 28 days.   Although it is impossible to eliminate any increase in long-term deflection due to early stripping it is generally accepted that as long as the first criteria above is met then any increase in deflection is likely to be less than 10 %, which is considered acceptable.

On the assumption that slab thicknesses range from 100 to 300 mm, Construction LL = 2 kPa and Design L.L = 3 kPa the first criteria yields a range for F_c / F'_c of 0.8 to 0.9.   The stripping times in AS3600 are based on a value of 0.85.

Looking at Table 9.1 a value of 0.85 is achieved at 24 days for 5 degrees, 18 days for 12 degrees and 12 days for 20 degrees which is in agreement with the figures in Table 19.6.2.5 of AS 3600.

9.9.4 Multifloor Formwork

Where two or more floors are formed to speed up cycle times the load of the wet concrete floor has to be supported by the floors below until they have hardened. Figure 9.10 illustrates the load in each floor expressed as a multiple of one floor dead load (eg 2D means that that particular floor has to carry twice its own dead load) as two sets of formwork proceed up the building. As will be illustrated below reshoring of recently poured slabs after they have gained sufficient strength to carry their own dead load results in a much reduced maximum load that has to be carried by a particular floor.

Without Reshoring

As the bottom forms are removed after the first pour the two slabs carry there own dead loads (1D). When the next floor is poured (3rd Pour) the dead load of the top floor must be carried equally by the two floors below (1.5D). When the bottom forms are then removed the top floor has sufficient strength to carry some of the extra load that was previously carried by the first slab (0.5D) – it shares this equally with the second slab making its load 0.25D and the second slab load 1.75D. Now when the 4th pour is poured the load of this slab is shared equally between the 2nd and 3rd floors giving a maximum load of 2.25D to be carried by the 2nd floor slab. At this stage the age of this floor is twice the cycle time.

With Reshoring

Reshoring involves relaxing the load on the props under a floor after it has gained sufficient strength to carry its own dead load. From the above second stage stripping criteria if we assume no construction load and dead load equal to live load then this will occur when the ratio of the concrete strength to the design strength is 1.2×DL/2.7×DL =0.44 and from Table 9.1 this would occur at three days after the pour for a temperature of 20°C.

In Figure 9.10 you can see that when the top slab is reshored after the 2nd pour it will carry its own load as will the 1st floor. After the 3rd floor is reshored it will carry its own load and the two slabs below will have their load reduced from 1.5D to 1D, following which the bottom set of forms will be removed. Now when the 4th floor is poured the load of that floor is shared equally between the 2nd and 3rd floors, giving a maximum load a slab has to carry of 1.5D. At this stage the weakest of the two floors carrying the load of 1.5D will be the 3rd floor which is only one times the cycle time old.

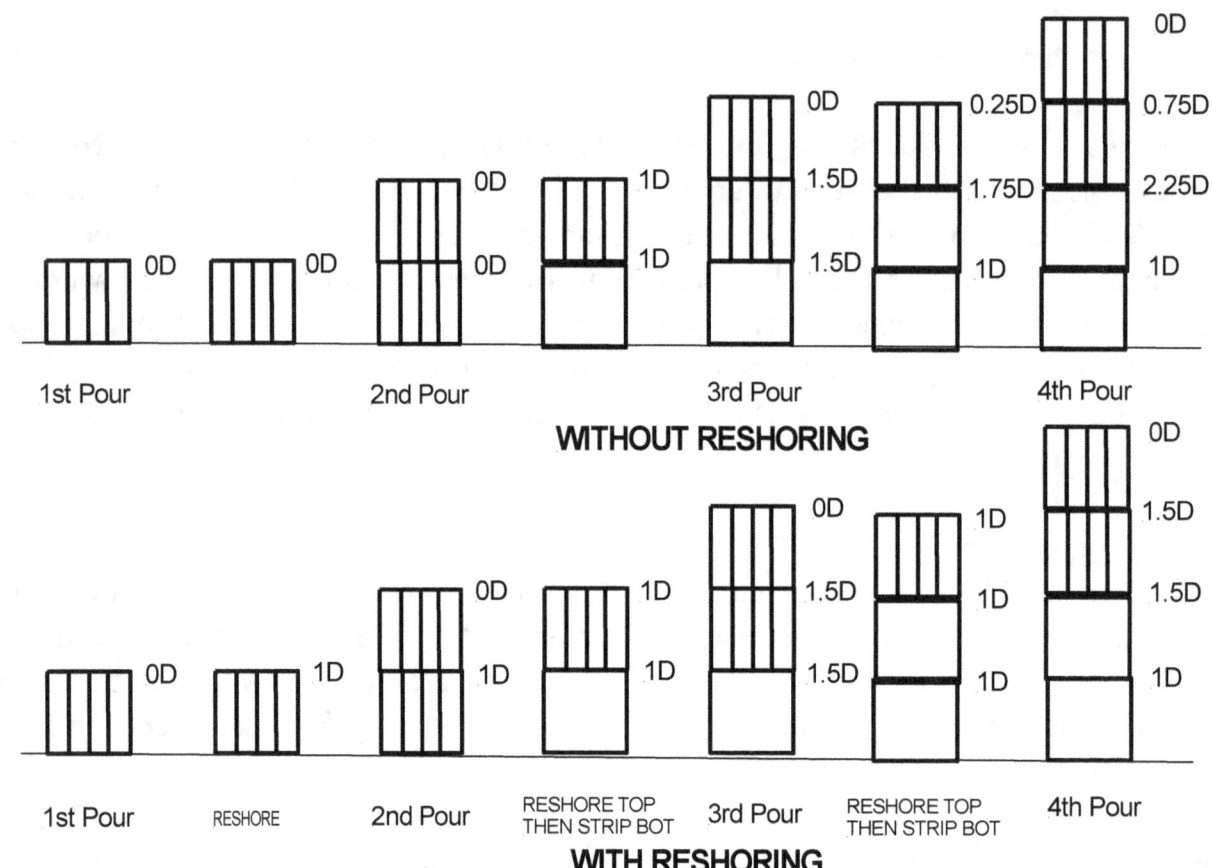

Figure 9.10 Multi-Story Stripping

The result is that with reshoring the maximum load to be carried by a slab is 1.5 D with the slab one cycle time old whereas without reshoring the maximum load to be carried is 2.75D with the slab being twice the cycle time old. Even though the slab in the reshored case is greener (less old) the benefit of the reduced load to be carried far outweighs the differences in ages of the slabs.

From above F_c/F'_c must be greater than
(1.2 × Construction D L + 1.5 × Construction L.L.) / (1.2 × D.L. + 1.5 × Design L.L)

If we assume no construction live load and dead load equal to design live load then F_c/F'_c must be greater than 1.2 × Construction D L / 2.7 × D.L. For the case of two sets of forms and reshoring the construction DL = 1.5 D from above (=1.5DL) then the requirement is that F_c/F'_c must be greater than 1.2×1.5/2.7 = 0.67. From Table 9.1 above this occurs at 6 days and therefore a cycle time of 6 days would be safe.

This analysis has not allowed for a construction live load on the slab being poured or for the dead load of the forms. These need to be allowed for and the critical slab then needs to be checked as above for second stage stripping

126

EXERCISES:

1.   Design a one way simply supported slab to span 4 metres carrying a live load of 3.3 KPa. Round up thickness to nearest 25 mm. Check deflection assuming SL82 mesh in the top of the slab. Assume Cover = 30 mm, F'c = 32 MPa, N12 bars.

2.   Design a three span continuous one way slab with spans of 5 metres carrying a live load of 4.65 KPa. Round up thickness to nearest 10 mm.   Check deflection assuming N12@350 in the compression zone of the end span. Cover = 20 mm, F'c = 25 MPa, N12 bars.

3.   Design a two way slab with four edges continuous having a span of 5 metres in one direction and 6 metres in the other. Assume a live load of 8.25 kPa, Cover = 25 mm, F'c = 32 MPa and N12 bars. Do not check deflection as it is not covered in this book. Draw a plan showing reinforcement.

4.   Work out the composite moment of inertia for a 80 thick slab sitting on top of a 250 UB31

5.   Assuming a 200 mm slab spanning 6.4 metres is propped in the centre determine the earliest time for stripping (F'c = 32 Mpa – Curing Temp = 5 ° C)
Ans: 4 days

6.   If F'c = 20 Mpa how long - AS 3600 says 7 days

7.   Assuming a design live load of 1.5 kPa, a construction live load of 0.5 kPa and a slab 150 mm thick determine the earliest stripping time if the curing temperature is 35 degrees centigrade.   Ans: 6 days

8.   What is the maximum slab load for three sets of forms and no reshoring.

9.   What is the maximum slab load for three sets of forms with reshoring.

10. BRACING OF HOUSE FRAMES

House frames need to be designed to resist the racking force of wind. The racking force is the force exerted on a particular face of a building which has to be resisted by bracing in walls at right angles to the face.   AS 1684 Timber Framing Code sets out the following method for determining the required bracing panels in houses.

10.1 Design Wind Categories

AS 4055-2006 Wind Load on Houses categorises wind conditions in terms of wind categories rather than wind speeds.   For other buildings the procedure given in Chapter 3 can be used to calculate wind pressures and racking forces. In order to find out the wind classification for a given location the following needs to be considered.

1. Wind Region.   Figure 2.1 of AS 4055-2006 Wind Load on Houses indicates regions for every part of Australia. Most of Australia is in Region A (Including Sydney) whilst Region B is coastal areas north of latitude 30 degrees (including Brisbane) and Region C is cyclonic areas.

2. Terrain Category. Terrain Categories are indicative of the roughness of the ground over which the wind blows. They range from Terrain Category 1 (Very Exposed open terrain), Category 2 (Open Terrain with well scattered obstructions such as around airfields), Terrain Category 2.5 (Terrain with a few trees and obstructions such as found in outer suburban areas) and   Terrain Category 3 (Terrain with numerous closely spaced obstructions such as houses in inner suburban areas)

3. Topographic Class. Houses on the flat are generally class T1 whilst houses on hills range from T2-T5 with T5 being the worst situation of a house at the top of a steeply sloping high hill.

4. Shielding Factor. Full shielding (FS) occurs when there are at least two rows of houses surrounding the house being considered. Other categories are partially shielded (PS) and No Shielding (NS)

Once these factors are known the wind classification is determined from Table 10.1 (Values taken from AS 4055)

Table 10.1 Wind Classification

Wind Region	Terrain Category	Topographic Class					
		T1			T2		
		Fully Shielded	Partially Shielded	No Shielding	Fully Shielded	Partially Shielded	No Shielding
A	3	N1	N1	N1	N2	N2	N2
A	2.5	N1	N1	N2	N2	N3	N3
A	2	N1	N2	N2	N2	N3	N3
A	1	N2	N3	N3	N3	N3	N4
B	3	N2	N2	N3	N3	N3	N4
B	2.5	N2	N3	N3	N3	N4	N4
B	2	N2	N3	N3	N3	N4	N4
B	1	N3	N4	N4	N4	N5	N5

10.2 Design Wind Pressures

AS 4055 gives tables for wind pressures on windward surfaces directly given that you know the Wind Class. These are given in Table 10.2 for Classes N1 to N3

Table 10.2 Wind Pressures on Windward Surfaces (kPa)

Wind Class	Gable End	Hip Roofs				
*incl. upper of double			Roof Pitch 20°		Roof Pitch 25°	
			Short	Long	Short	Long
N1	0.66	Single*	0.52	0.45	0.56	0.52
		Double	0.60	0.52	0.61	0.58
N2	0.92	Single*	0.72	0.62	0.77	0.72
		Double	0.84	0.72	0.86	0.81
N3	1.44	Single*	1.10	0.95	1.20	1.10
		Double	1.30	1.10	1.30	1.30

For single storey buildings use the "Single" rows. For double storey buildings the "Single" rows apply to the bracing requirement in the upper storey and the "Double" values apply to the bracing requirements in the lower storey. "Short" means wind pressure on short wall.

10.3 Converting Wind Pressures to Racking Forces

Racking forces are the forces due to wind on windward surfaces <u>in the direction of the wind</u>. So for example if wind is hitting the short end of a building the calculations for racking force determine the bracing required in the long walls, i.e. walls at right angles to the wind. These may be internal or external walls. To counteract racking forces braces are required in these walls eg metal angle bracing ("SpeedBrace") or plywood sheeting.

To calculate the force required to be braced we need to calculate the force on the projected area of the surfaces <u>perpendicular</u> to the direction of wind. Figure 10.1 shows the areas of a building to be considered in calculating the racking force.

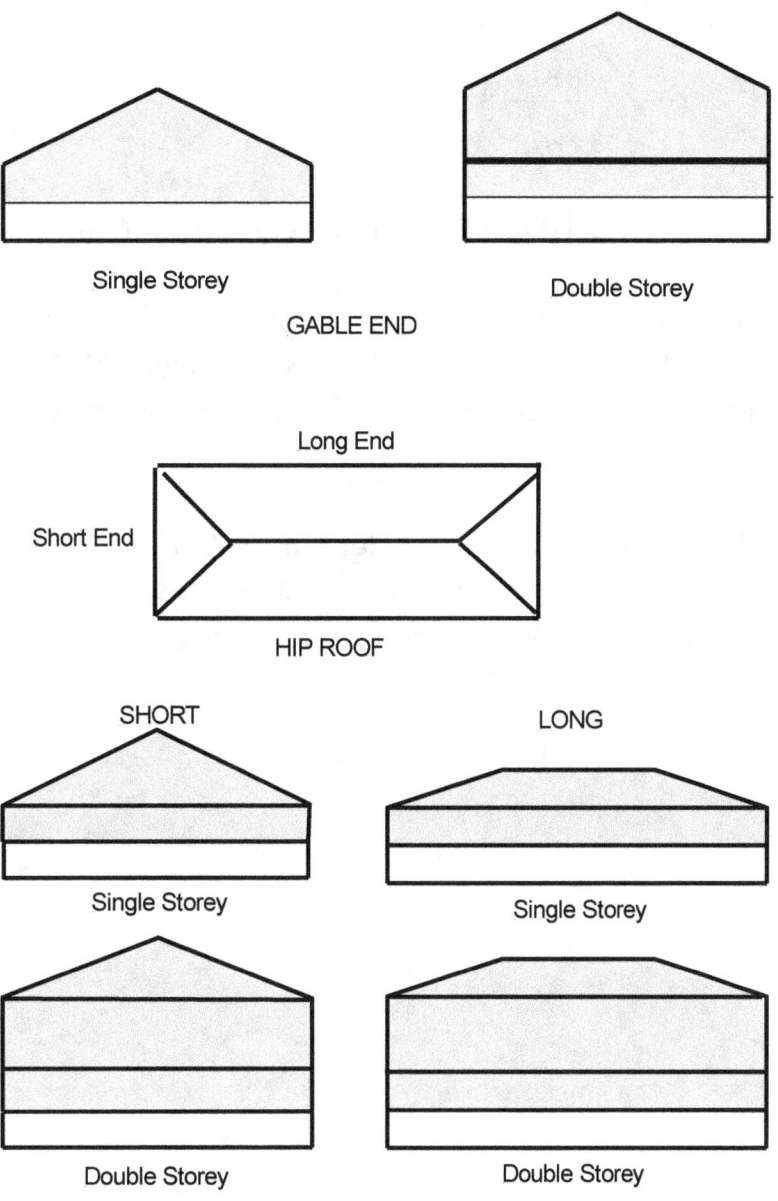

Figure 10.1   Projected Areas for Racking Force Calculations

The racking force is then determined as

$$\text{Racking Force (kN)} = \text{Area (m}^2) \times \text{Wind Pressure (kPa)}$$

10.4 Choosing Required Bracing

Once the racking force has been calculated we can choose the type and size of bracing required. Brace capacities of various length of walls cross braced with 30 by 0.8 mm gal straps are given in Figure 10.2.

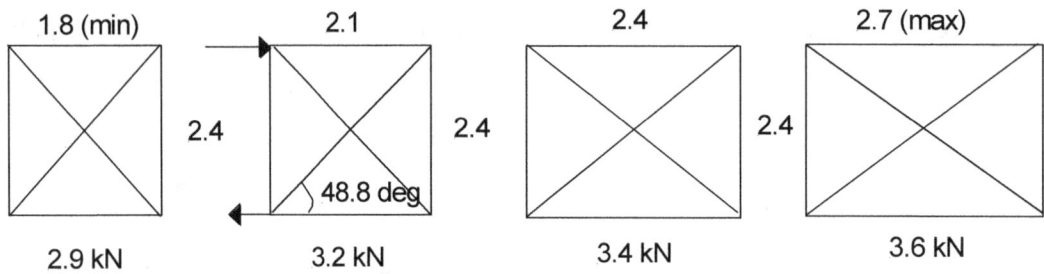

Figure 10.2 Capacities of Various Lengths of Braced Walls

To calculate how many braces of a particular length are required divide the total racking force by the relevant brace capacity given in Figure 10.2. Note that the code requires a minimum of two braces in each external wall and that internal walls may also be braced.

Note that Australian Standard 1684 "Residential Timber Frame Construction" gives a nominal value of 1.5 kN/m of bracing which is conservative for the shorter braces but slightly unconservative for the longer braces.

10.5 Racking Force Example

Consider a hip roof building 12 m long by 6 m wide with a roof pitch of 20° in a Class N1 area.

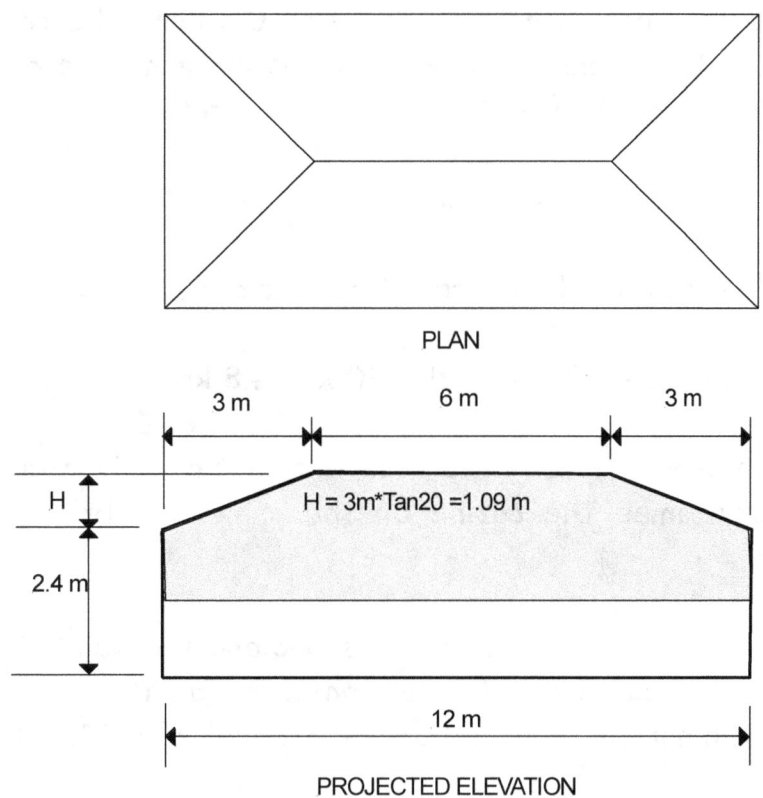

PLAN

H = 3m*Tan20 =1.09 m

PROJECTED ELEVATION

Bracing requirements for the short walls are calculated as follows.

From Table 7.1 wind pressure = 0.45 kPa (wind on long wall)

Projected Area = 12×1.2 +9×1.09 = 24.21 m^2
Racking Force = 24.21 × 0.45 = 10.9 kN
Minimum 4 braces required (2 in each wall)

If 1.8 m long braces were used we would need 10.9/2.9 = 3.75 (4) braces

Say 4 / 1.8 m long double braces on the short side, one in each corner.

10.6 Calculation of Brace Capacities

Values given in Table 10.2 are calculated as follows.

Consider the case of cross- bracing with 30 by 0.8 mm gal straps fastened with 3.15 mm dia nails. The capacity of such straps in tension is given by the same equation as for timber members but without the k_1 factor.

$$\phi N_t = \phi \times A_n \times F_t / 1000$$

Assuming the yield stress of the brace is 250 MPa and $\varnothing = 0.9$ gives

Brace Capacity = $0.9 \times [(30-3.15) \times 0.8] \times 250/1000 = 4.8$ kN

The portion of the brace capacity available to resist a horizontal force is equal to the brace capacity times the cosine of the angle the brace makes with the horizontal.

If for example the brace panel is 2.1 metres long and the wall height is 2.4 metres the angle the brace makes with the horizontal is $\text{Tan}^{-1}(2.4/2.1) = 48.8°$ and the brace capacity in the horizontal direction is $4.8 \times \text{Cos}48.8° = 3.2$ kN

Note that in AS1684 cross braces are given a capacity rated in KN/m length of brace (1.5 kN/m). For example a 2.1 metre length of braced wall would have a capacity of $2.1 \times 1.5 = 3.2$ kN (same as in Figure 10.2) whilst a 2.7 metre long braced section would have a capacity of $2.7 \times 1.5 = 4.0$ kN (slightly higher than in Figure 10.2).

EXERCISES:

1. A two storey house is in Region A with Topographic Class 2 and Terrain Category 2. It is partially shielded.

The building is 10 metres by 5 metres in plan, ceiling height is 2.5 metres and the roof pitch is 25 degrees.

10 m

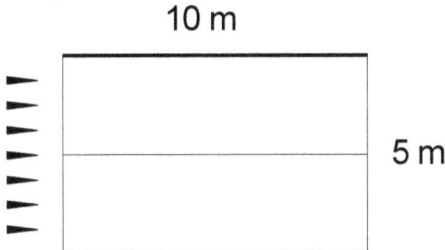

5 m

Calculate the bracing required on the long side in the lower storey due to wind force on the <u>gable</u> end.

2. A two storey house is located in wind region B, T.C. 2.5, Topographic Class T1 and is fully shielded.

The building is 6 metres by 4 metres in plan, ceiling height is 3 metres and the roof pitch is 20 degrees.

6 m

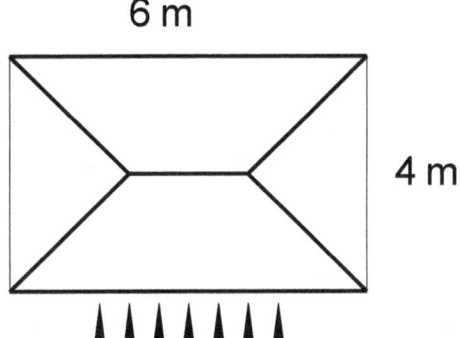

4 m

Calculate the bracing required on the short side walls in the lower storey due to wind force on the <u>long side</u>.

11. LATERAL LOAD RESISTING STRUCTURES

11.1 Introduction

Wind loads are the primary load to be resisted by industrial buildings. Wind loads are also the primary consideration for multi storey buildings and in some cases earthquake loads may be significant. Both wind and earthquake loads are "lateral" loads on buildings.   The calculation of wind load pressures has been covered elsewhere in these notes. Earthquake loads can be approximated by an equivalent horizontal static load equivalent to a proportion of the total vertical dead load of the structure. This proportion typically is between 5 and 20 %.

11.2 Lateral Load Resisting Elements

11.2.1   Cross Bracing

Steel bracing elements capable of carrying tension or compression or both are common forms of stabilising structures against lateral loads. In its simplest form one might consider the guy wires of a tent or the stays on a sailing boat.

For buildings cross bracing can take many forms, as illustrated below

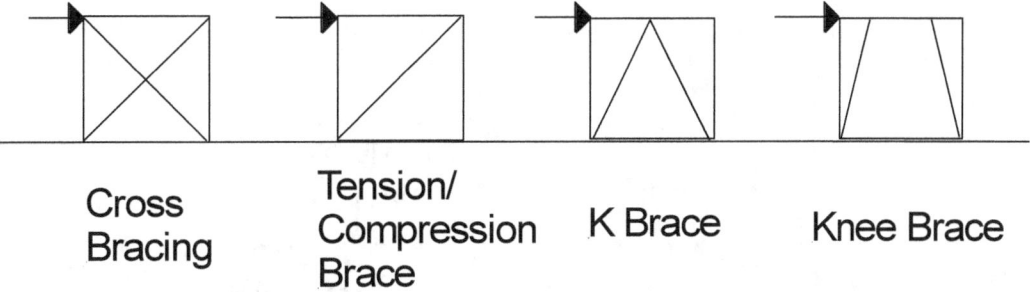

| Cross Bracing | Tension/ Compression Brace | K Brace | Knee Brace |

The bracing for the first three of these act in tension or compression whilst in the case of the knee bracing the floor member acts in bending as well as tension/compression.

The knee bracing solution is the least effective, as can be seen from the following deflection diagrams, but it offers the greatest flexibility in placing openings within the facade.

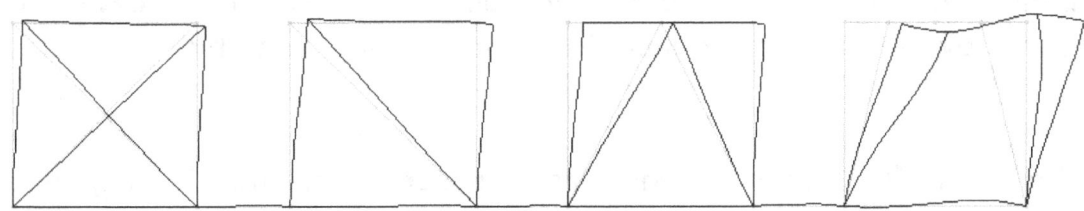

11.2.2 Shear Walls

SHEAR WALL

Shear walls resist lateral forces by their shear resistance. Looking at the wall above it is easy to see how a lateral load at the top is transferred to the bottom through the shear resistance of the bricks and mortar joints, the latter being the weak link.

Shear walls behave essentially as cantilever beams and, although they are called "shear" walls, their mode of deformation is one of "bending"

Shear walls take many forms in multi-storey buildings. They can be the dividing walls between rooms or units in the case of residential buildings, or a central slip formed core.

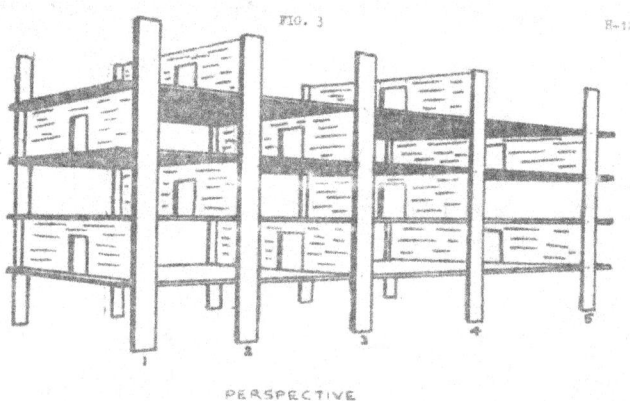

PERSPECTIVE

11.2.3 Rigid Frames

Rigid frames occur commonly as "portal frames" in industrial buildings but they also can occur in multi-storey buildings. Reinforced concrete frame buildings by their very nature have rigid joints but rigid joints can also be achieved in steel construction by welding beam/column joints or by using high strength friction grip bolts.

Rigid frames deflect in a "shear" mode under horizontal loading. and have significant bending moments in both columns and beams (See Figure 11.1)

The diagram below shows the deflection and bending moments of a portal frame subjected to a horizontal load at the top of the left hand side column.

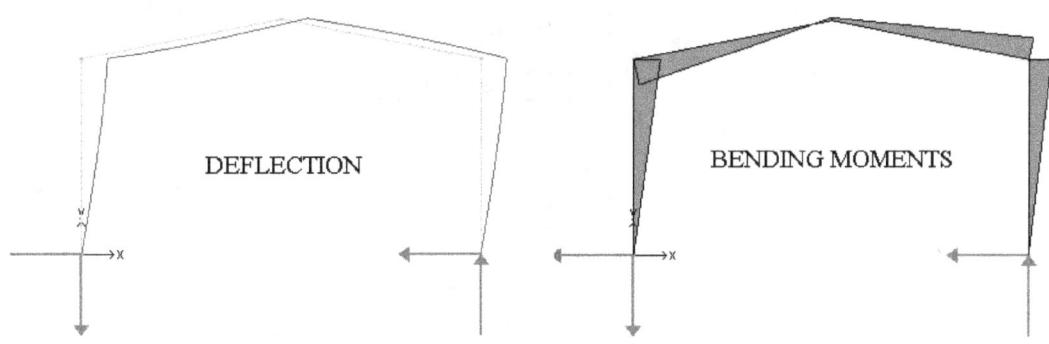

DEFLECTION   BENDING MOMENTS

11.2.4   Cantilever Tubular Structures

Tubular lateral load resisting structures (think of a flagpole) are commonly utilised in multi-storey structures. Slip formed concrete cores in multi-storey buildings behave essentially the same as flagpoles and resist deformation by virtue of their stiffness, which is a function of the moment of inertia of their cross-section.
In some cases eg Twin Towers in New York the tube is made up of closely spaced columns on the periphery of the building. This is much more efficient as it significantly increases the moment of inertia of the cross-section.

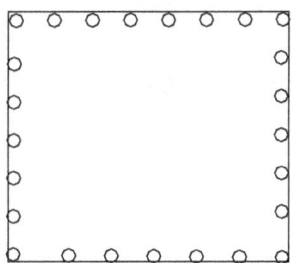

Because the side walls are stiff the corner columns tend to carry more tension/compression than the face columns;

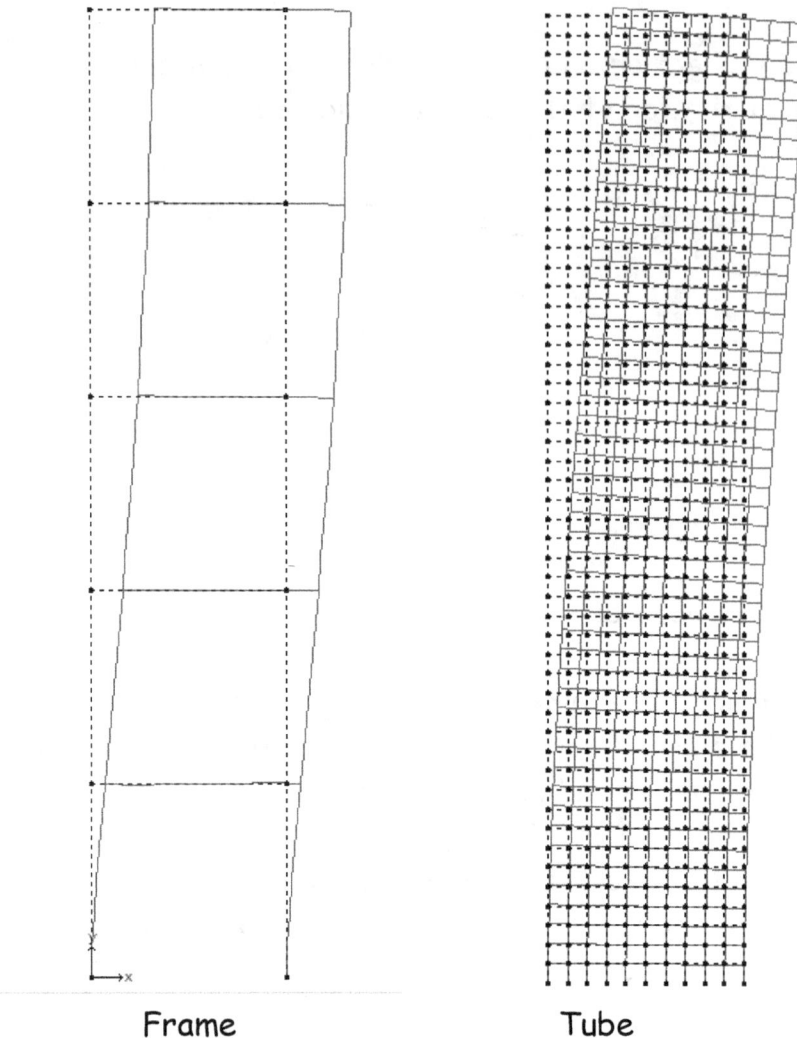

Frame   Tube

Figure 11.1 Relative Deflection Profiles of Frame vs Tubular Structure

11.3 Structural Systems

11.3.1   Portal Frames

Portal frame buildings resist lateral loads in one direction by rigid frame action (portal frame) and in the other direction by trusses in the roof plane which connect to diagonal bracing in the walls.

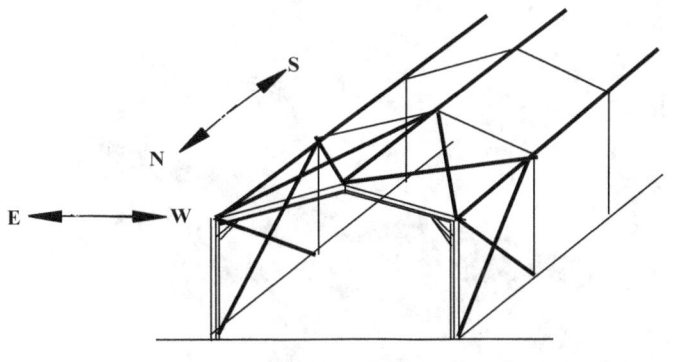

E-W WIND RESISTED BY PORTAL FRAME
N-S WIND RESISTED BY CROSS-BRACING

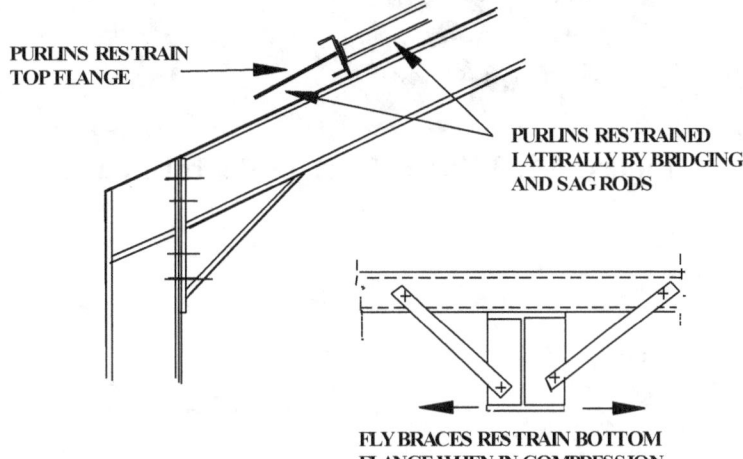

PURLINS RESTRAIN
TOP FLANGE

PURLINS RESTRAINED
LATERALLY BY BRIDGING
AND SAG RODS

FLY BRACES RESTRAIN BOTTOM
FLANGE WHEN IN COMPRESSION

Figure 11.2 Cross bracing of Portal Frame

141

Figure 11.3 Knee Joint in Portal Frame

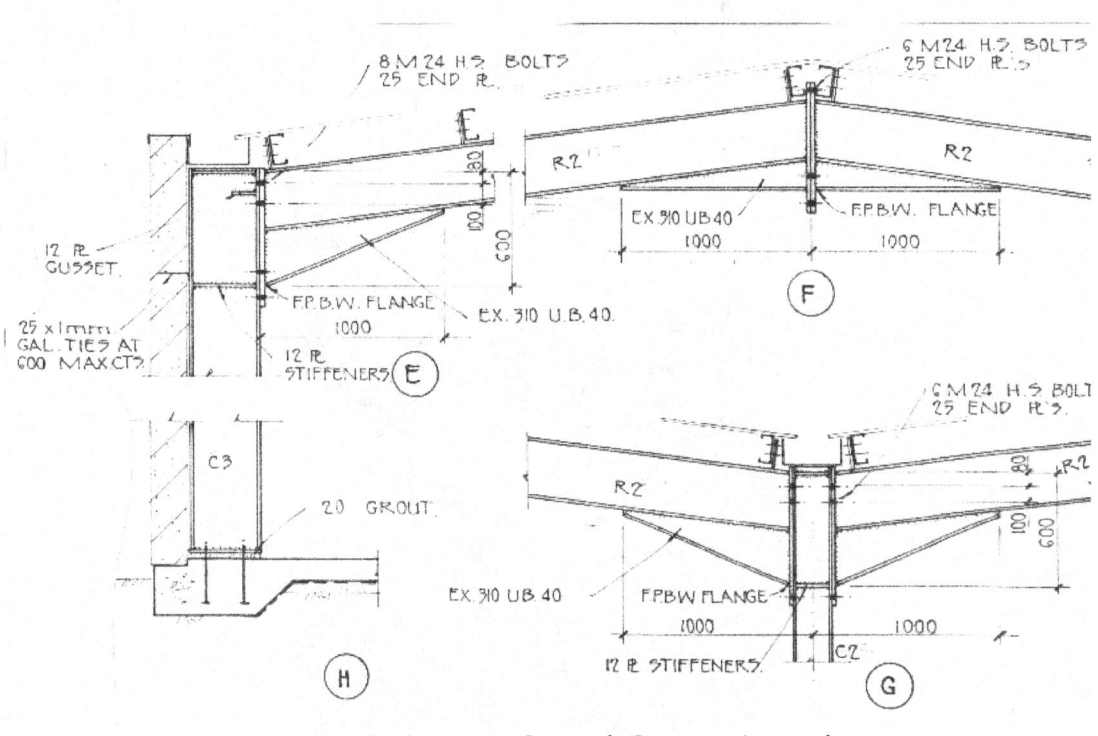

Figure 11.4   Portal Frame Details

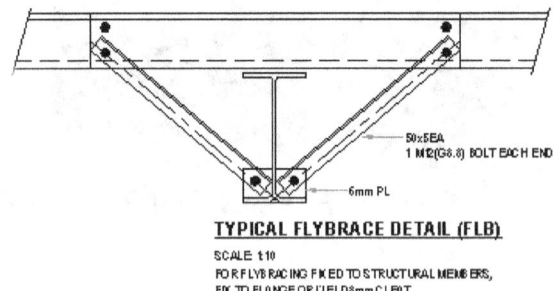

Figure 11.5   Fly Brace Detail

142

11.3.2   Rigid Frame Multi-storey Buildings

These rely on frame action to resist loads and are generally limited to buildings less than about 15 storeys
Multi-Storey rigid frame buildings subjected to horizontal loads and uniformly distributed floor loads have deflections and bending moments similar to those shown below

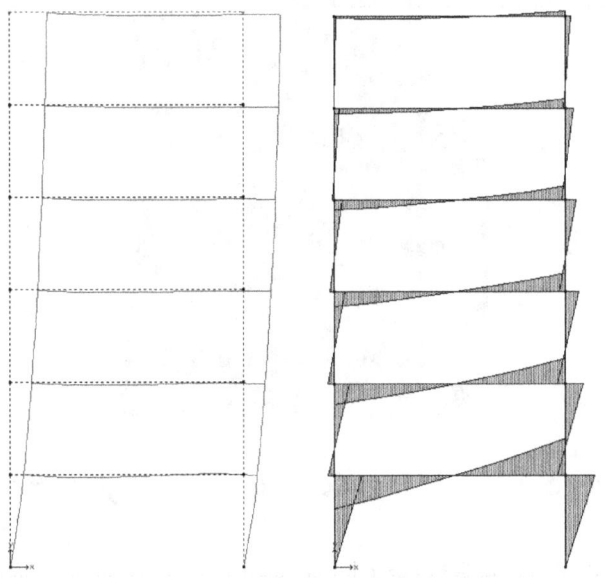

Deflections   Bending Moment Diagram

With sophisticated frames incorporating girder trusses heights of up to 50 storeys are possible, as in the Hong Kong and Shanghai Bank building in Hong Kong.

Figure 11.7 Hong Kong and Shanghai Bank

11.3.3   Shear Wall Multi-Storey Buildings

These typically have a slip or jump formed core which carries all the lateral loads. The exterior columns carry vertical load only. Such buildings are only suitable up to around 20 storeys.

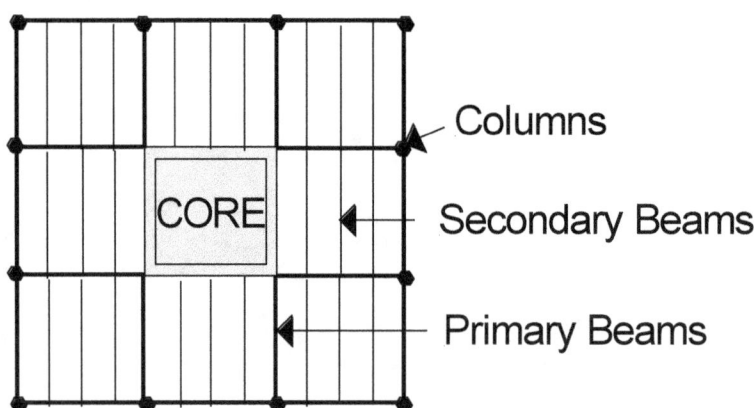

11.3.4   Shear Walls with Interacting Frame Multi-Storey Buildings

Shear wall cores can be combined with frame action to provide a stiffer structure with buildings up to 100 storeys.

Outrigger systems are a modified form of braced frame and shear wall frame systems. They comprise a central shear wall core which is connected to the external columns at frequent intervals by trusses which are generally one or two storeys tall. Taipei 101 is an example of this system.

Figure 11.8 Taipei 101

11.3.5   Tube Multi-Storey Buildings

These are buildings with closely spaced columns on the exterior which make the building behave as a tube. The World Trade Center Towers were construction in this manner. If required the internal core can be utilised to resist lateral loads (tube in tube) but this requires expensive moment connections between tubes

Figure 11.9 Twin Towers, N.Y. (Prior to 911)

11.3.6   Bundled Tube Multi-Storey Buildings

Although quite an old concept, as illustrated by its use in the 108 story Sears Tower in 1974, the bundled tube offers flexibility of architecture by allowing specific bundles to be stopped off at various floor levels. This building remained the tallest building in the world until 1996.

11.3.7   Braced Tube Multi-Storey Buildings

The example of this type of construction is the Bank of China building in Hong Kong. By adding bracing to the closely spaced external columns the shear lag effect is much reduced.

11.3.8   Worlds Tallest Building

The Burj Dubai is currently the tallest building in the world. Because the building is essentially residential it makes use of shear walls combined with the tapering efficiency of its "Y" shaped plan to resist lateral loads. It is not a bundled tube building like Sears Tower.

Figure 11.7 Bank of China Tower (Hong Kong)

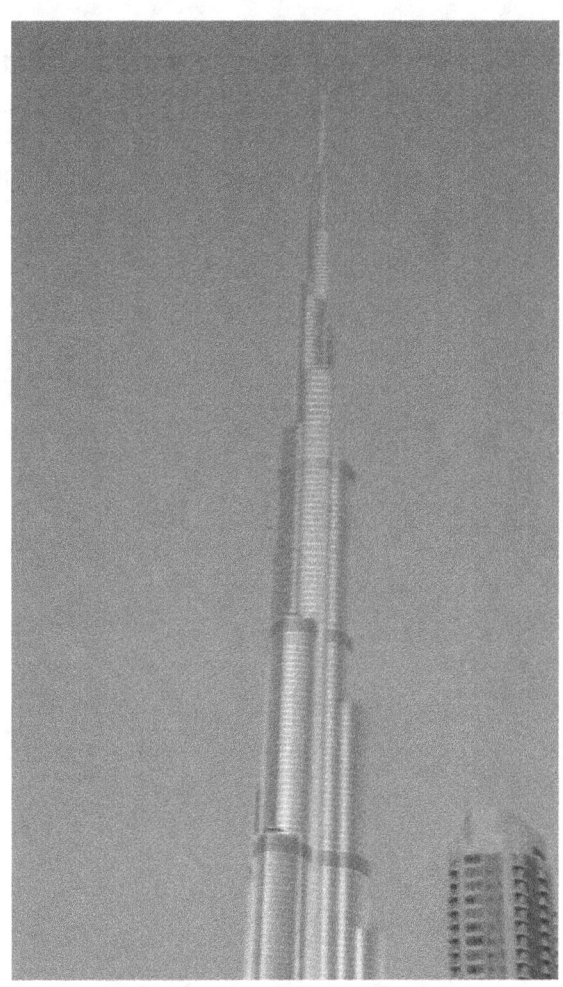

Figure 11.8 Burj Dubai

11.4 Transfer Floors and Beams in Multi-Storey Buildings

These are required to reduce the number of columns and open up foyer areas.

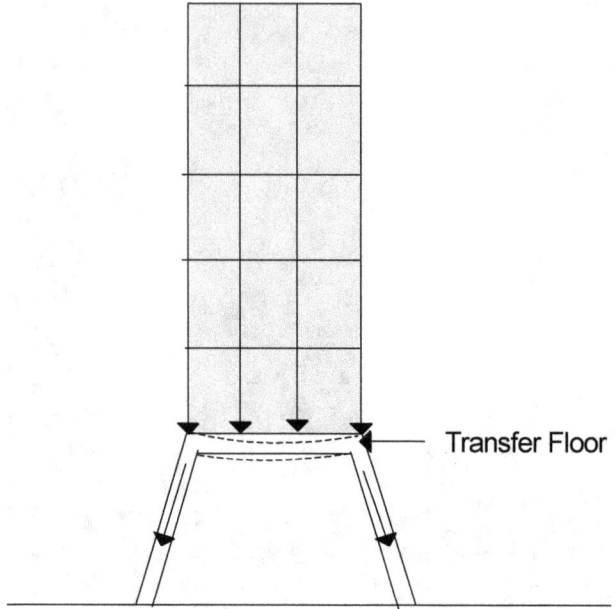

Transfer Floor

11.5 Mass Dampers

These are used to limit lateral movement due to wind or earthquakes.

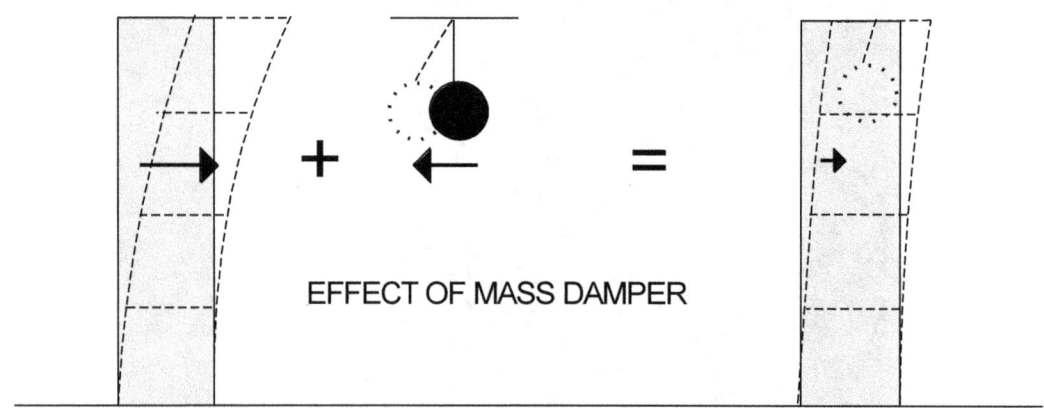

EFFECT OF MASS DAMPER

Appendix A. USEFUL MATHEMATICS AND SECTION PROPERTIES

A.1 Mathematics

1. $1 \times 10^3 = 10x^y3 = 1000 = 1EXP\ 3$
2. $10 \times 10^5 = 1,000,000 = 1 \times 10^6 = 1EXP\ 6$ or $10EXP\ 5$
3. $2 \times 10^2 = 200$ not 20^2
4. $\frac{45 \times 3}{10 \times 2} = \frac{135}{20} = 6.75$ or $(45 \times 3) \div (10 \times 2)$
5. $25 \times 10^6 \times 4 = 100,000,000 = 1 \times 10^8 = 1EXP8$
6. $14,280,000 = 1.428 \times 10^7 = 1.428EXP7$
7. $\frac{5}{384} \times \frac{6000 \times 1000^3}{10,000 \times 10 \times 10^5} = \frac{5 \times 6000 \times x^y3}{384 \times 10,000 \times 10EXP5} = \frac{3 \times 10^{13}}{3.84 \times 10^{12}} = 7.8$
8. $(0.22 \times 6) + 0.73 = 2.05$ NOT $1.48 = 0.22 \times 6.73$
9. $-40 + 20 + 4F1 = 0$

   $-20 + 4F1 = 0$

   $4F1 = 20$

   $F1 = 5$

   [-60+30-(4×2)-3F2=0 Gives F2 = -12.67]
10. To solve $1.4 \times e^{-0.1}$ first $e^x(-0.1) = 0.9$ then times by 1.4 to get 1.27 (e is naperian log- e^x button on calculator)

    [1.4×e(-0.1×10) = ex(-1)×1.4 = 0.5]

A2. Geometry

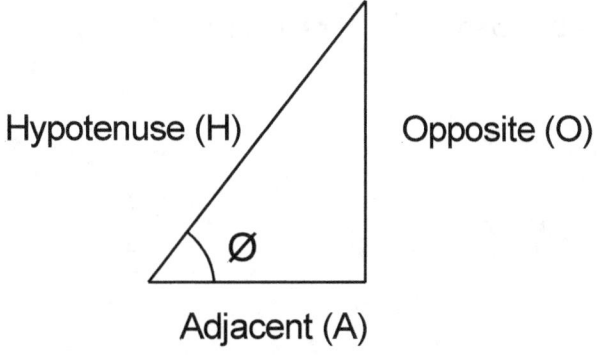

Make sure you are in degrees mode on your calculator, not radians

Sine Ø = Opposite/Hypotenuse
Cos Ø = Adjacent/Hypotenuse
Tan Ø = Opposite/Adjacent

Opposite = Adjacent × Tan Ø

A.3 Section Properties

1. Rectangle

Area = b×d (b = width, d = depth in direction of load)

Moment of Inertia (I_{xx}) = b×d³/12

Section Modulus (Z_{xx}) = b×d²/6

Radius of Gyration (r_y)= 0.29 b d

2. Circle   Area = π×d²/4 (d = diameter)

Moment of Inertia = 0.049×d⁴

Section Modulus = 0.0982×d³

Radius of Gyration = 0.25×d

3. Circular Tube

Area = π×(d_o^2-d_i^2)/4

(d_o = outside dia., d_i = inside dia.)

Moment of Inertia = 0.049×(d_o^4-d_i^4)

Section Modulus = 0.0982× (d_o^4-d_i^4)/d_o

Radius of Gyration = 0.25×(d_o^2+d_i^2)$^{0.5}$

4. Triangle

Area = 0.5×b×h (b = width, h = perpendicular height)

Moment of Inertia (I_{xx}) = b×h³/36

Minimum Section Modulus = (Z_{xx}^{top}) = b×h²/24

Distance from bottom to Centroid = h/3

5. Cone or Pyramid

Volume = Base Area×h/3 (h = perpendicular height)

Distance from bottom to Centroid = h/4

6. Sphere

Volume = 0.5236×d³ (d = diameter)

Appendix B. TYPICAL LIMIT STATE CAPACITIES OF STRUCTURAL MEMBERS

B.1 Compressive Capacity of Common Members

(NOTE: THESE ARE FOR CENTRALLY LOADED MEMBERS ONLY)

Member	Compressive Capacity (kN)			
	l = 1 m	l = 2 m	l = 3 m	l = 4 m
100 by 50 F7 Oregon*	10	2	-	-
100 by 100 F7 Oregon*	55	19	9	-
100 by 100 F11 Hwd*	86	28	13	7
150 UB 14	335	171	86	50
100 UC 15	397	291	175	109
150 PFC	589	352	198	122
76 by 5.9 CHS	275	218	135	81
100 by 4 CHS	268	241	193	135
100 by 100 by 5 SHS	552	488	374	252
50by 50 by 4 SHS	180	88	42	24
No 2 Acrow Prop**	-	51	20	-
No 3 Acrow Prop**	-	56 (2.2 m)	24	12
No 4 Acrow Prop**	-	-	30 (3.1 m)	17

* k_1 =0.8, Effective sizes 90× 45, 90×90, unseasoned,φ = 0.65.

** Values obtained from Acrow Data multiplied by 1.5 to get φN_c

B.2 Typical Bending Moment Capacities

Timber Beams (Restrained Compression Flange φ =0.65, k1 =1.0 (Actual Sizes))

Beam	Moment Capacity (kN.m)
150 by 50 F7 (140 × 45)	1.9
200 by 50 F7 (190 ×45)	3.5
200 by 75 F7 (190 × 70)	5.5

Steel Beams (Unrestrained Compression Flange

Beam	Moment Capacity (kN.m)				
	l= 2 m	l = 3m	l = 4 m	l = 5 m	l = 6 m
150 UB18	21.6	17.0	13.9	11.7	10.0
180 UB 22	33.3	26.5	21.7	18.2	15.7
200 UB25	50.5	41.1	33.4	27.6	23.3
200 UB30	61.9	51.5	43.0	36.3	31.2
250 UB 31	78.8	64.6	52.3	42.8	35.7
150 PFC	23.3	19.4	16.4	14.1	12.4
200 PFC	39.3	32.0	26.7	22.7	19.7
100UC15	14.3	12.4	10.8	9.6	8.5
150 UC 23	37.2	32.2	27.6	23.9	20.9

Reinforced Concrete Beams (Restrained Compression Flange)

Beam	Moment Capacity (kN.m)
300 wide by 350 deep, 2Y24 B	75
200 Blockwork, Y16@400	21 / metre

Appendix C.   SECTION PROPERTIES OF STRUCTURAL ELEMENTS

C.1   Cross Sectional Area (A)

This is generally used for calculating compressive or tensile stresses in a member eg column. For normal rectangular members of width "B" and depth "D" $A = B \times D$.

C.2   Centroid

The centroid is the centre of the cross sectional area and is used for calculating Moment of Inertia and Radius of Gyration. Its position is measured from some reference axis (usually bottom left hand point of section) with co-ordinates y_b and x_b. For normal rectangular members the centroid is in the centre.

C.3   Moment of Inertia (I)

The moment of inertia of a section reflects the ease with which it bends (given similar materials). The more material that is located away from the centroid the higher the moment of inertia. Moments of inertia can be different about the x-x axis (I_{xx}) compared to the y-y axis (I_{yy}). Generally we are concerned with I_{xx}. For normal rectangular members of width "B" and depth "D" $I_{xx} = B \times D^3/12$ and $I_{yy} = B^3 \times D/12$.

C.4   Radius of Gyration (r)

The radius of gyration of a section measures its resistance to buckling. Typically we are concerned with buckling about the weak y-y axis in which case we are interested in r_{yy}. For normal rectangular members of width "B" and depth "D" $r_{yy} = 0.289 \times B$. For other sections $r_{yy} = \sqrt{(I_{yy}/A)}$ and $r_{xx} = \sqrt{(I_{xx}/A)}$

C.5   Section Modulus (Z)

The section modulus of a beam reflects its strength in bending. As bending is normally about the x-x axis we are usually interested in Z_{xx}. For normal rectangular members of width "b" and depth "d" $Z_{xx} = B \times D^2/6$ and $Z_{yy} = B^2 \times D/6$.

The section modulus about an axis can be different for the top of the beam (Z_{xx}^{top}) than the bottom of the beam (Z_{xx}^{bot}), but generally we use the lower of the two values.
$Z_{xx}^{top} = I_{xx}/y_{top}$ and $Z_{xx}^{bot} = I_{xx}/y_{bot}$, where y_{top} and y_{bot} are the distance from the top and bottom to the centroidal x axis.

C.6   Calculation of Y-Y Centroidal Axis (X-X Similar)

Step 1: Divide section into reactangles.
Step 2: Calculate areas of rectangles and total area (A)

Step 3: Take moments of individual rectangle areas about reference axis (usually at bottom of section (side for X-X). If A_1 is area of rectangle 1 and y_1 is the distance from the centre of the rectangle to the reference axis then $M_1 = A_1 \times y_1$
Step 4: Distance of centroid of section to reference base y_b is then equal to the sum of the moments divided by the total area ($\sum M / A$)

Example:

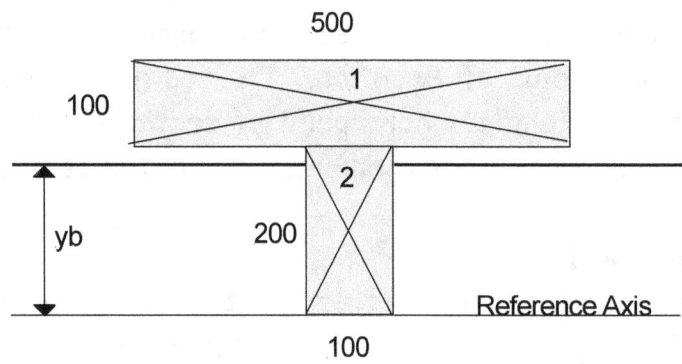

Area 1 = 500×100 = 50,000
Area 2 = 200×100 = 20,000
A= 70,000
M_1 = 50,000×250 = 12.5×10^6
M_2 = 20,000×100 = 2.0×10^6
$\sum M$ = 14.5×10^6
y_b = 14.5×10^6/70,000 = 207

C.7   Calculation of I_{xx} (I_{yy} Similar)

Step 1: As for calculating centroidal axis
Step 2: For each rectangle calculate its moment of inertia ($I = B \times D^3 / 12$)
Step 3: For each rectangle calculate its moment of inertia about the x-x axis ($I^* = A \times d^2$, where d is the distance of the centre of the rectangle from the calculated X-X axis).
Step 4: Sum all the above moments of inertia to get total Moment of Inertia of Section about X-X axis (I_{xx}).

Example using Section above:

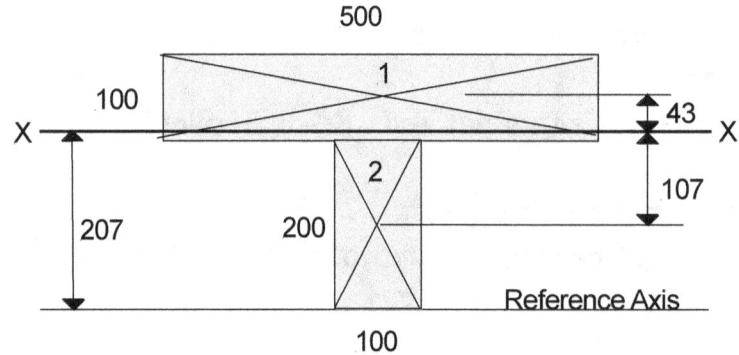

154

$I_1 = 500 \times 100^3 / 12 = 41.7 \times 10^6$

$I_2 = 100 \times 200^3 / 12 = 66.7 \times 10^6$

$I^*_1 = 50{,}000 \times 43^2 = 92.5 \times 10^6$ (Note $d_1 = 43 = 300-207-50$)

$I^*_2 = 20{,}000 \times 107^2 = 229.0 \times 10^6$ (Note $d_2 = 107 = 207-100$)

Then Moment of Inertia of Section = $\sum I = 429.9 \times 10^6$ (mm^4 if dimensions in mm)

C.6 Steel Section Properties

Section (Depth by Mass/m)	Depth (mm)	Flange Width (mm)	Flange Thickness (mm)	Web Thickness (mm)	Area A (mm2)	Moment of Inertia I_{xx} (mm4 $\times 10^6$)	Section Modulus Z_{xx} (mm3$\times 10^3$)	Radius of Gyration r_{yy} (mm)
150UB14	150	75	7.01	5.00	1790	6.67	88.9	16.7
150 UB18	155	75	9.50	5.99	2300	9.05	117	17.1
180UB18	175	90	8.00	5.00	2310	12.1	139	20.6
180UB22	179	90	10.00	5.99	2820	15.3	171	20.8
200UB25	203	133	7.82	5.84	3230	23.6	232	31.0
200UB30	207	134	9.60	6.30	3800	28.9	279	31.8
250UB31	251	146	8.64	6.10	4000	44.4	353	33.5
250UB37	256	146	10.90	6.40	4750	55.6	434	34.7
310UB40	304	165	10.20	6.10	5150	85.2	561	38.5
310UB46	307	166	11.80	6.73	5890	99.5	648	39.0
360UB45	352	171	0.73	6.86	5700	121.0	687	37.8

EXERCISES:

Calculate all section properties for the following sections

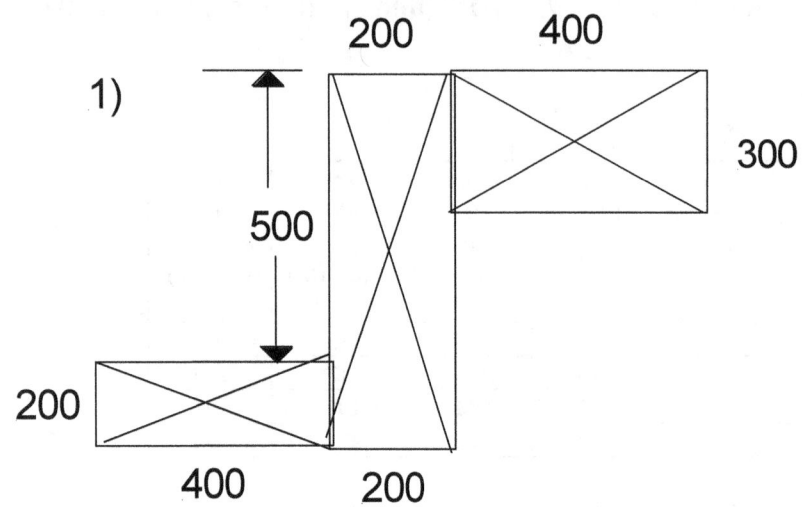

1)

Approximation of 180 UB 22

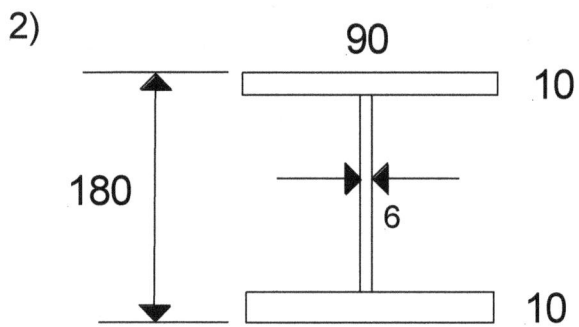

2)

Answers

1.   ryy = 246.8 mm

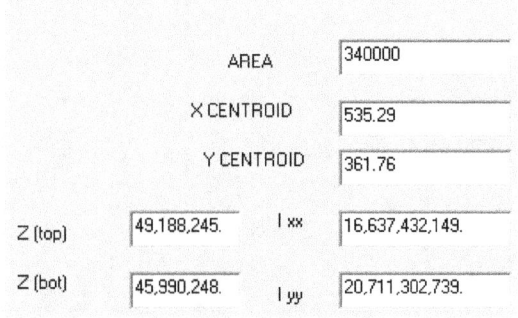

AREA	340000	
X CENTROID	535.29	
Y CENTROID	361.76	
Z (top)	49,188,245.	I xx	16,637,432,149.
Z (bot)	45,990,248.	I yy	20,711,302,739.

2   Check with Section C.6

156

Appendix D   DETERMINATION OF FORCES IN FRAMES USING GRAPHICAL METHOD

Consider the case of the following steel frame subjected to the wind loads shown. Reactions are calculated using the three equations of equilibrium.

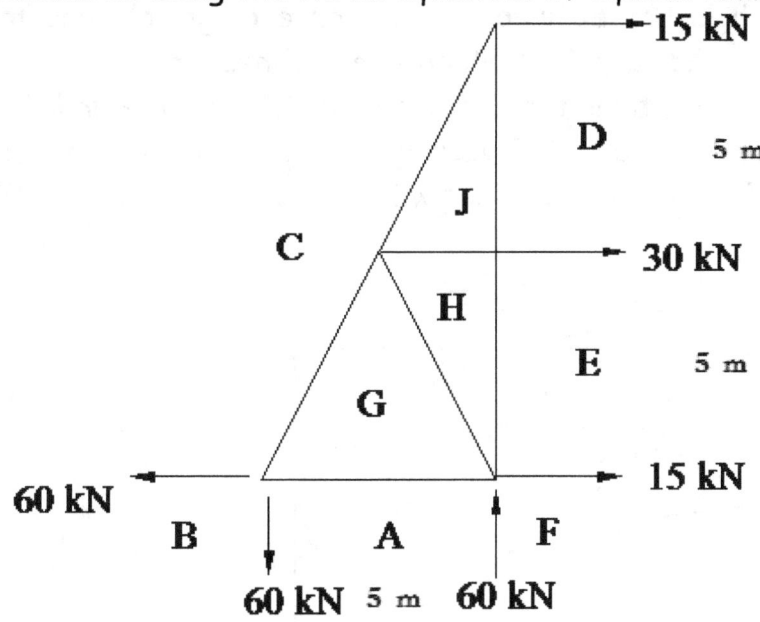

Loading Diagram

The first step is to label areas between external loads and internal areas. Areas are labelled clockwise alphabetically between external loads and then internal areas are labelled. Define members in the following way. Member A-G is the member between areas "A" and "G". Now start off by marking the point "A" on a fresh sheet of paper. Mark this point in the middle to give you room and define a scale in terms of kN say 1 cm = 10 kN. Moving clockwise around the diagram follow the arrows to define points "B", "C", "D", "E" and "F". The final move should bring you back to point "A".

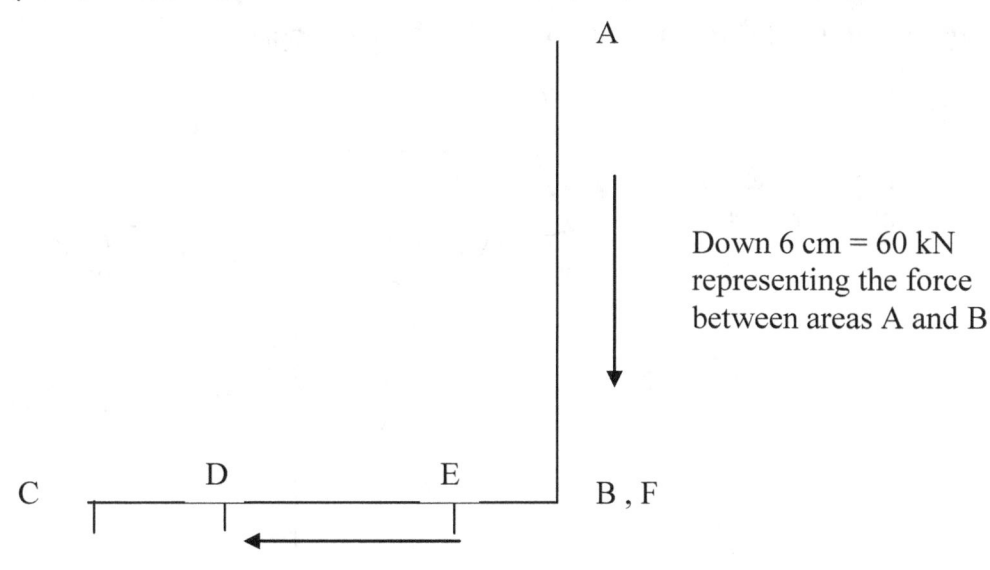

First Step in Drawing Force Diagram

Now on the <u>loading diagram</u> start at any one of the nodes where there is one surrounding area which is not defined on the force diagram (only one not two). In the example the left hand support node has area "G" surrounding it and point "G" is not defined on the force diagram , so we will start here.

Draw lines parallel to members containing unknown points (eg Member A-G contains the point "G" which is not defined on our force diagram) and make them pass through the corresponding points which are defined on the force diagram. For example Member A-G contains the defined point "A" and the undefined point "G" so draw a line parallel to Member A-G passing through point "A" on the force diagram. The intersection of these two new lines will define the new point ("G" in this case)

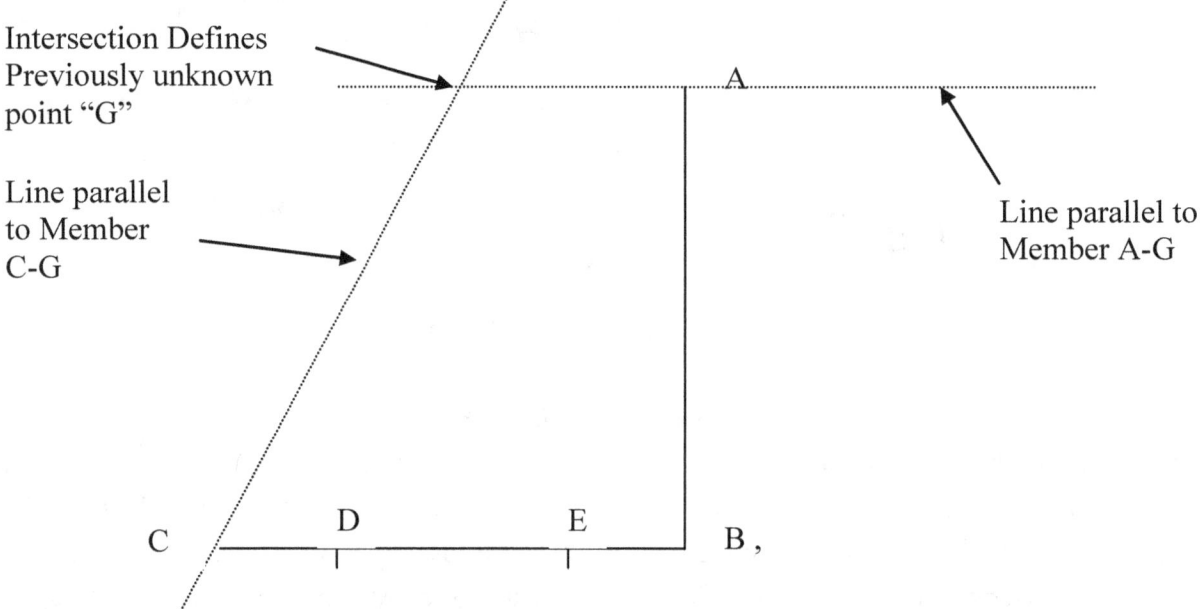

Having now defined the point "G" we can now scale off the magnitude of the forces in members A-G and C-G. Scaling them off the diagram we get Force A-G = 30 kN and force C-G = 67 kN. We can now proceed to the next node where there is only one undefined point. In the above case we move to the node at the right hand support where point "H" is undefined on the force diagram.

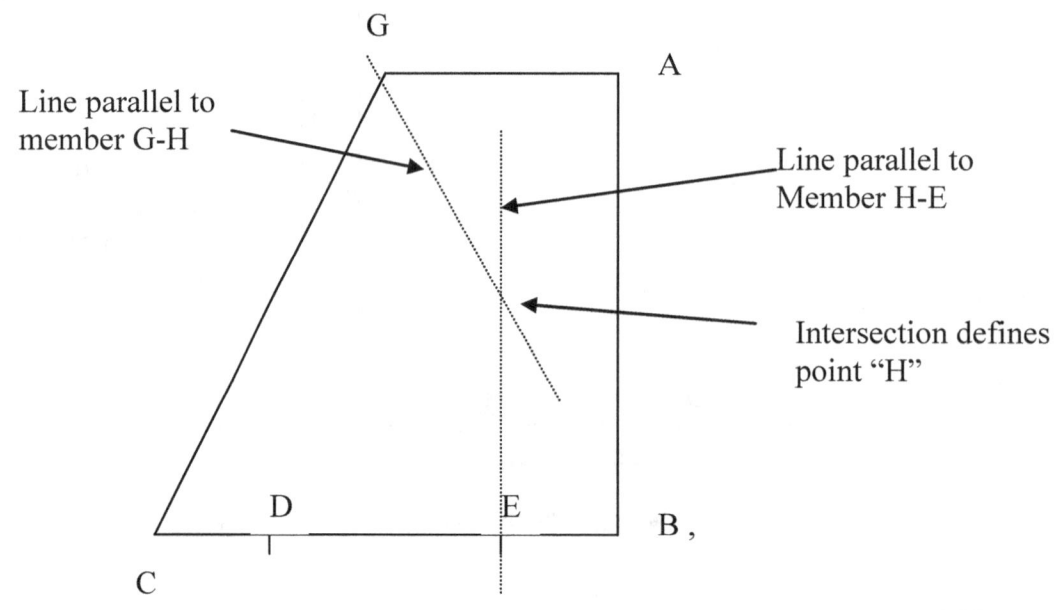

158

The complete force diagram is then as follows

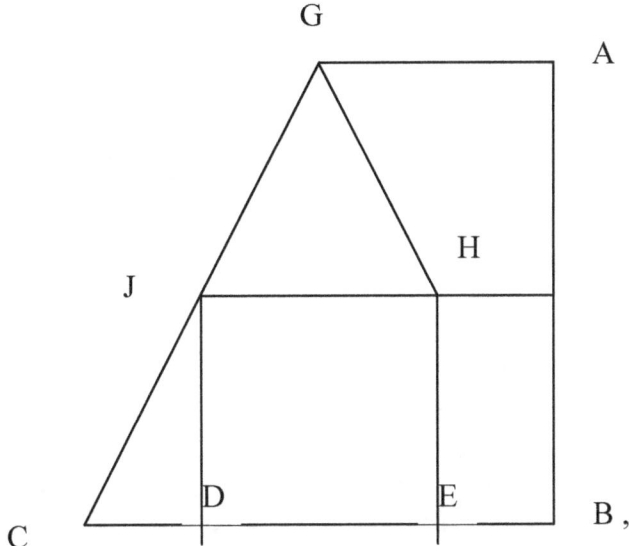

All the forces can now be scaled off the force diagram. To figure out the sign of a force i.e. Tension (+ve) or Compression (-ve) hold your finger on a joint where the member intersects (Say the left hand support joint). Move clockwise around this joint. On rotating clockwise around this joint you move from area "A" to "B" to "C" to "G" and back to "A". For member C-G you are moving from area "C" to area "G". Now go to the force diagram and move similarly i.e. from point "C" to point "G". You will notice that you are moving in a north-easterly direction. Go back to the joint and draw an arrow on member C-G in a north-easterly direction. If this arrow moves away from the joint then member C-G is in tension (Towards the jpoint would be compression. In this case the arrow moves away from the joint so the force in Member C- G is in tension i.e. +ve.

The forces in the truss are shown below with compression forces negative.

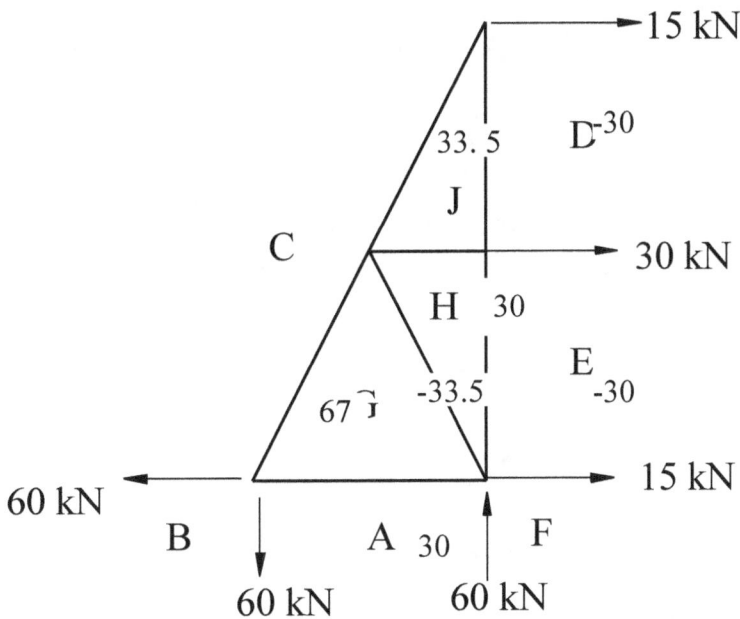

Appendix E.   ANSWERS TO EXERCISES

Chapter 1:
1a)   27.5 kN
1b)   87.1 kN
1c)   3.06 kN
1d)   6.93 kPa
2.   0.52 kPa
3.   6.5 kPa
4.   72 kN
5.   DL =0.75 kPa, LL=0.87 kPa, Total Load Pressure = 1.62 kPa
6.   0.97 kN/m

Chapter 2:
1.   2.9 kPa
2. P_{max} = 18.3 kPa, F = 36.72 kN/m acting 1.33 metres from base.
3.   0.34 kPa inwards

Chapter 3:
1.   Firm Sandy Clay. ABP = 100 kPa
2. B_e = 225 mm   Allowable Load = 9 kN
3.   Be = 0.3 m   Pressure =147 kPa < 150kPa therefore OK
4.   N = 12 at 2.5 m giving ABP = 180 kPa,
      e/B = 0.167 m, B_e = 0.4 m, Pressure = 167 kPa so OK
5.   H = 45.4 kN, W = 235.2 kN, e/B = 0.093, B_e = 1.855, Pressure =127 kPa <225

Chapter 4:
1a) V_L = 20 kN, V_R = 20 kN, M_{max} =20 kN.m

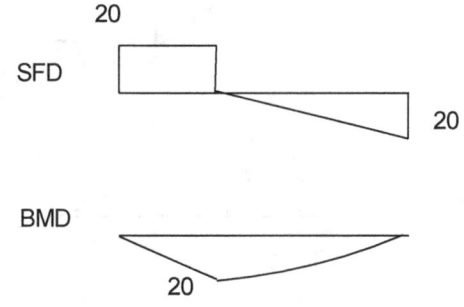

1b) V_L = 19.5 kN, V_R = 31.5 kN, -ve M_{max} =20 kN.m , +ve M_{max} = 23 kN.m

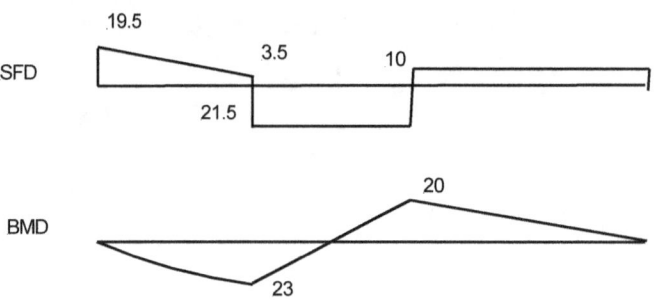

1c) V_L = 38 kN, V_R = 32 kN, M_{max} =65 kN.m

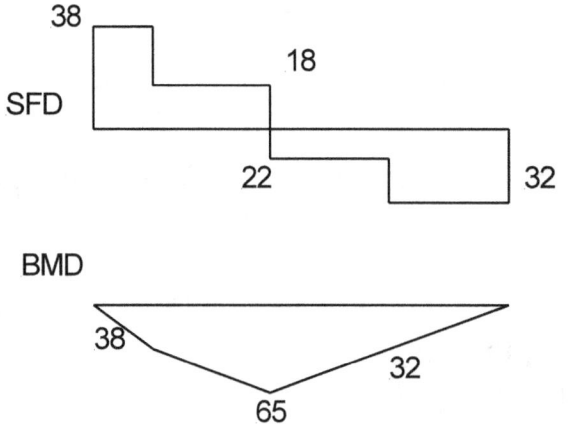

1d) V_L = 6.25 kN, V_R = 48.75 kN, -ve M_{max} =26.25 kN.m , +ve M_{max} =1.95 kN.m

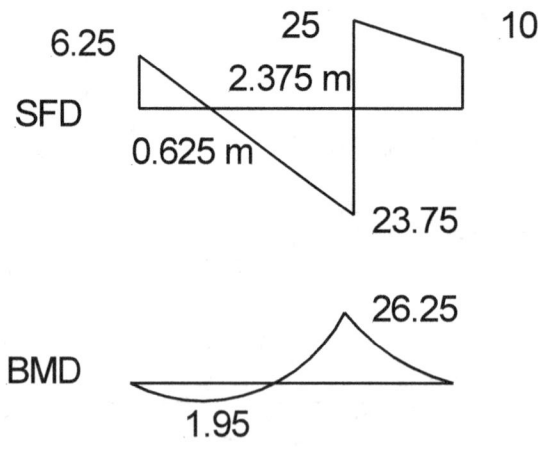

1e) V_L = 33.33 kN, V_R = 36.67 kN, M_{MAX} = 37.78 kN.m

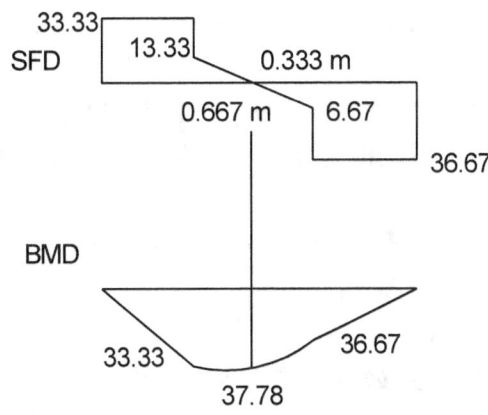

2.   Top Chord 61.3 kN, Bottom Chord 58.7 kN.

3.

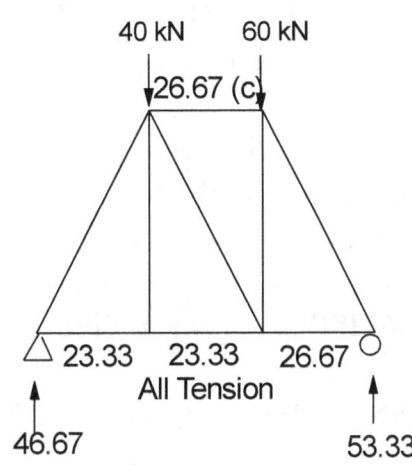

4.   Deflection = 10.24 mm

Chapter 5:
1. $M_{O/T}$ =1.26 kN.m M_{RES} =1.68 kN.m
2. $M_{O/T}$ = 1736 kN.m (Wind +DL×1.2) = 1728 kN.m (DL×1.35)
        M_{RES} = 1890 kN.m (2.5 m thick).

Chapter 6:
1. M^*=22.65 kN.m, $\emptyset M_b$ = 65.5 kN.m OK, V^* = 7.56 kN, $\emptyset V$ = 76.3 kN OK,
        δ(long term) = 56 mm < 60 mm Therefore OK for deflection.
2.   170 by 35 seasoned F7 @ 600
        Strength (k_9 = 1.13)
        Case 1 M^* = 0.58kN.m $\emptyset M_b$ = 1.82 kN.m OK
        Case 2 M^* = 1.24kN.m $\emptyset M_b$ = 2.56 kN.m OK

Deflection

Case 1 δ = 4.58 mm δ(allowable) = 8 mm   OK

Case 2 δ = 1.27 mm δ(allowable) = 2 mm   OK

3.   Continuous 190 by 45 seasoned F7 @ 450

     Strength (k_9 =)

     Case 1 M^* = 1.22 kN.m $\emptyset M_b$ = 3.11 kN.m   OK

     Case 2 M^* = 2.58 kN.m $\emptyset M_b$ = 4.37 kN.m   OK

     Deflection

     Case 1 δ = 6.2 mm δ(allowable) = 13 mm   OK

     Case 2 δ = 2.4mm δ(allowable) = 2 mm   Not OK but OK if 3 spans

4. N^* = 18 kN, $\emptyset N_t$ =20 kN   OK

5. N^* = 6.75 kN, k_{12} = 0.039, $\emptyset N_c$ = 7.2 kN OK

Chapter 7:

1. M^* = 100 kN.m, k_{12} = 0.77, a_m = 1.33, $\emptyset M_b$ = 115 kN.m,

     δ = 13.1 mm, $\delta_{allowable}$ =15 mm

2. M^* = 10.8×w, l_e = 4400 , k_{12} = 0.57, $\emptyset M_b$ = 127 kN.m , δ = 32 mm

3.   Deflection (200UB 25) δ = 15.2 mm, $\delta_{allowable}$ =16.7 mm

     Strength (250UB 37) M^* = 94.9 kN.m,

     k_{12} = 0.435, a_m = 2.25 , $\emptyset M_b$ = 109.9 kN.m

4.   250 UB 31. δ = 13.9 mm, $\delta_{allowable}$ =20 mm

     M^* = 46.4 kN.m, k_{12} = 0.8, a_m = 1.0, $\emptyset M_b$ = 73 kN.m

5.   d = 458 mm, M^* = 283.5 kN.m, Ast = 1795 mm^2 , 4/N24 bars, p=0.6%

Creep factor 2.17 (=3-1.2×1256/1809),   w for deflection = 34.2 kN/m

δ long term = 19 mm, $\delta_{allowable}$ =20 mm

6.   d = 408 mm, M^* = 280 kN.m, Ast = 1990 mm^2 , 5/N24 bars, p=.82%

Creep factor 2.47 (3-1.2×1005/2260), w for deflection = 33.5 kN/m, δ long term =

30 mm, $\delta_{allowable}$ =20 mm.   Therefore 10 mm precamber required.

7.   d = 290 mm, M^* =79.5 kN.m, Ast = 795 mm^2, 3/N20 bars, p=1%

w for deflection = 19.4 kN/m, Creep Factor = 2.49 (3-1.2×402/942), δ long term =

19 mm, $\delta_{allowable}$ =13 mm. No good- either precamber 6 mm or make beam deeper

Chapter 8:

1. M^* = 0.092 KN.m/m, ϕM = 0.098 kN.m/m   Therefore OK...

Chapter 9:

1.   D= 200 mm, d=164 M^* = 21.9 kN.m/m, Ast = 354mm^2/m, N12@300 B1

w_{LT} = 6.1kPa, I = 1.456E8 mm^4/m, E= 28,595, CF = 2.2, δ = 10.7 mm, δ(allowable)

=13.3 mm

2.   D =180mm, d = 154mm, M-ve = 31 kN.m/m, Ast = 533mm^2/m, N12@200 top.

End span M+ve = 28.2 kN.m/m, Ast = 485mm^2/m, N12@225 Bottom.

Interior Span M+ve = 19.5 kN.m/m, Ast = 335mm^2/m, N12@325 Bottom.

A_{sc} = 113/0.35 = 323 mm^2/m, A_{st} = 113/0.225 = 502 mm^2/m , Creep Factor = 2.23

w_{LT} = 6.05 kPa ,I = 1.205E8 mm^4/m, δ = 15 mm, δ(allowable) =16.7 mm

3.   D= 150 mm, dx = 119, dy = 107, w* = 16.88 kPa

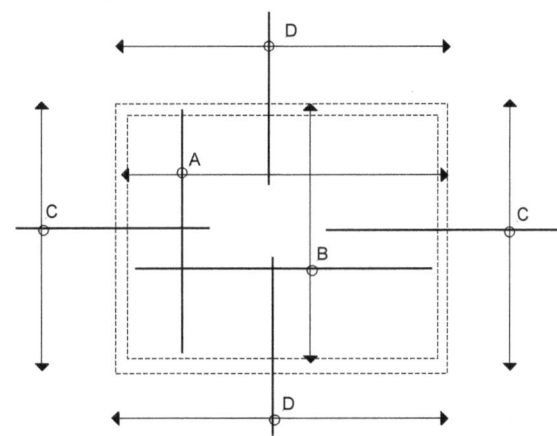

A N12@370 Bot(L1), B N12@400 Bot(L2), C N12@320 Top(L3), D N12@280 Top(L4), (L1 means laid first etc)

4. y_b =233 I_{xx} = 119.8×106 mm^4
5.   4 days
8.   AS 3600 says 7 days
9.   6 days
10. 2.36D
11. 1.33D

Chapter 10:

1.   N3 Classification, A = 21.68 m^2, Force = 31.22 kN, 10/2.1 metre braced lengths (Note some will need to be in internal walls)
2.   N2 Classification, p = 0.72 kPa, A = 29.92 m^2, Force = 21.6 kN, 4/1.8 metre braces in external walls and 6/2.1 metre braces in internal walls

www.ingramcontent.com/pod-product-compliance
Lightning Source LLC
Chambersburg PA
CBHW081055170526
45166CB00006B/2078